IS ALL WE KNOW A BLACK HOLE?

IS ALL WE KNOW A BLACK HOLE?

CARL TYSON

First Printing, 2025

CONTENTS

Introduction

The universe, in its silent majesty, presents us with a spectacle of unparalleled beauty and profound mystery. From the delicate dance of galaxies across cosmic distances to the enigmatic whispers of the earliest moments of creation, we are constantly presented with phenomena that challenge our understanding and ignite our deepest curiosity. For generations, humanity has looked to the heavens, seeking to unravel the secrets of our cosmic home. We have built intricate models, conducted meticulous observations, and engaged in vigorous debate, all in an effort to piece together the grand tapestry of existence. Yet, despite our advancements, the universe continues to hold its breath, revealing just enough to tantalize, but never quite enough to satisfy our insatiable hunger for knowledge.

This book embarks on a journey into one of the most radical and mind-bending propositions in modern cosmology: the idea that our universe, the entirety of spacetime and matter we know, might have originated from within a black hole. This is not a casual speculation but a concept that emerges from the rigorous, albeit often speculative, landscape of theoretical physics. It challenges our most fundamental assumptions about beginnings and endings, about confinement and expansion, and about the very nature of reality itself. We will navigate through the well-established principles of Einstein's General Relativity, probe the strange anatomy of black holes, and reconsider the familiar narrative of the Big Bang. By drawing parallels between the extreme physics governing these cosmic entities and the observed characteristics of our universe, we aim to explore this hypothesis with both intellectual rigor and a sense of profound wonder. Prepare to question what you thought you knew, for in the pursuit of cosmic truth, the most extraordinary ideas often lie just beyond the edge of our current understanding.

The human endeavor to comprehend the cosmos is as old as consciousness itself. From ancient mythologies depicting celestial spheres to the sophisticated mathematical models of modern astrophysics, we have consistently strived to map our place within the grand, unfolding narrative of the universe. We observe the breathtaking expanse, the intricate structures, and the subtle interplay of forces, all of which point towards a profound and underlying order. Yet, at the heart of our cosmological understanding lie profound mysteries, enigmas that have persisted through decades of scientific inquiry. How did it all begin? What is the fundamental nature of the forces that shape reality? And could our familiar universe, with its expanding horizons and enigmatic dark components, be more intricately connected to cosmic phenomena we once considered simply destructive forces?

This book ventures into the speculative, yet scientifically grounded, territory of a universe born from within a black hole. This proposition, while audacious, is not born from fantasy, but from the persistent questions and theoretical challenges that arise when we push the boundaries of our current cosmological models. We will begin by establishing a firm foundation in our current understanding of the universe, including the triumphs of the Big Bang theory and the elegance of Einstein's General Relativity, which describes gravity as the curvature of spacetime. We will then delve into the anatomy of black holes – cosmic entities of such immense gravitational power that they warp spacetime to its breaking point, featuring singularities and event horizons that remain among the most enigmatic features of the cosmos. It is at this intersection of extreme gravity, quantum mechanics, and cosmic origins that the radical hypothesis we explore finds its conceptual roots. By examining the theoretical pathways that connect the properties of black holes to the observed characteristics of our universe, we aim to present a compelling case for why this idea deserves serious scientific consideration. This journey will take us from the familiar cosmos to the heart of theoretical physics, seeking to illuminate the possibility that the birth of our universe might be intimately linked to the very phenomena we once perceived as the ultimate cosmic end.

CHAPTER 1

The Unseen Embrace: A Universe Within

Our universe, a vast expanse of galaxies, nebulae, and the enigmatic void, has always captivated humanity and been a focus of scientific investigation. From ancient stargazers to modern instruments probing the farthest reaches of space, our quest to comprehend our cosmic home has defined our civilization. We have mapped constellations, uncovered stellar life cycles, and developed theories explaining the universe's birth, evolution, and destiny. Yet, despite these achievements, the cosmos remains deeply mysterious, hinting at phenomena and structures beyond our current understanding.

It is within this blend of knowledge and enigma that we introduce a radical concept: the possibility that our entire observable universe exists within a black hole. This idea is not a proven fact but a compelling theoretical frontier, born from the convergence of established physics and speculative thought. It invites us to explore new intellectual territory, questioning our cosmic origins and the seemingly endless space we inhabit. This initial exploration aims to ignite curiosity and encourage an open-minded consideration of our familiar universe as a unique, internal realm, shaped by forces we are just beginning to grasp.

To fully appreciate this hypothesis, we must first ground ourselves in our current, widely accepted understanding of the cosmos. We trace our journey from the Big Bang—a state of unimaginable density and temperature—through rapid expansion and cooling. We have observed the formation of subatomic particles, the coalescence of atoms, and the gravitational dance that created stars, galaxies, and the vast cosmic web. The discovery of the cosmic microwave background radiation (CMB), a faint echo of that primordial fiery state, supports the narrative of a universe that began explosively and has been expanding and cooling ever since. We understand the universe as a dynamic entity, constantly evolving, shaped by gravity, matter, and energy across an immense canvas. This established model, refined over decades of observation and theory, forms the bedrock against which we will contrast the unconventional idea of a universe contained within a black hole. Establishing this baseline is crucial to fully grasp the radical nature of the proposition we are about to explore.

Black holes, traditionally seen as cosmic anomalies or destructive forces, may hold a more complex and generative role in cosmology. Initially perceived as inescapable traps of immense gravitational power, crushing matter and light into an infinitely dense singularity behind an event horizon, a more nuanced picture is emerging. These intense gravitational forces, the enigmatic singularity, and the definitive event horizon are not solely features of cosmic destruction, but also potent subjects for radical cosmological theories. This shift in perspective suggests that black holes, far from being mere cosmic vacuum cleaners, could be cradles of new realities—the very wombs from which entire universes might be born.

The genesis of such a seemingly outlandish idea—that our universe could reside within a black hole—stems from the frontiers of our current theoretical understanding. It arises from unexplored avenues and persistent theoretical inconsistencies in existing cosmological models, rather than a void of scientific reasoning. Puzzles like the ultimate origin of the Big Bang singularity or the nature of spacetime at extreme densities naturally lead to unorthodox explanations. This hypothesis is not an isolated spark of imagination but the culmination of an intellectual lineage that consistently pushes scientific boundaries. Throughout history, radical cosmological theories have emerged when existing paradigms face insurmountable obstacles, sometimes paving the way for revolutionary breakthroughs despite initial skepticism. Outlining this intellectual journey highlights the scientific courage and curiosity driving such exploration, tracing the idea's evolution from established physics to a compelling theoretical frontier.

The significance of exploring this unconventional hypothesis cannot be overstated. Dedicating intellectual resources to an idea that might initially seem fantastical is valuable for advancing scientific understanding. Such explorations act as intellectual crucibles, testing current theories and revealing their limitations. This specific hypothesis, for instance, could offer novel explanations for persistent cosmological questions. It may provide a framework for unifying disparate cosmological observations, offer a new perspective on the enigmatic origin of the Big Bang singularity, and potentially alter our understanding of reality, causality, and existence itself. Venturing into these speculative realms expands our scientific knowledge and deepens our appreciation for the universe's vastness and complexity. This chapter concludes by emphasizing the exciting, uncharted territory ahead, inviting the reader to explore this truly enigmatic possibility.

The hypothesis that our universe exists within a black hole, while seemingly a product of science fiction, stems from established physics, particularly Einstein's theory of General Relativity and the evolving field of quantum gravity. These theories, especially within extreme gravitational contexts, provide the foundation for such radical ideas. A black hole, defined by spacetime so warped that light cannot escape, represents a theoretical limit where our conventional understanding of physics breaks down. The event horizon, this fundamental spacetime feature, results from an immense concentration of mass and energy, marking the point where reality is maximally stretched and distorted—the inception point of our hypothesis.

The central question concerns the fate of matter and energy crossing the event horizon. General Relativity posits that all infalling matter is drawn to a singularity at the black hole's core, compressed to infinite density. However, considering quantum reality, particularly at the Planck scale where quantum gravitational effects are expected, the singularity's nature becomes less certain. Theories like Loop Quantum Gravity or String Theory suggest the singularity might not be a point of infinite density, but rather a region of extremely high, finite density where spacetime could undergo a quantum "bounce," shifting from extreme compression to rapid expansion.

This "bounce" concept critically links black hole physics to the Big Bang. If a black hole's singularity, under quantum gravity, leads not to an end but to a rebound, then conditions within a black hole could, in principle, generate a new, expanding universe. Imagine matter and energy, highly compressed at the singularity, being violently expelled or reforming into a rapidly expanding state. This outward surge of energy and spacetime would strongly resemble the Big Bang. In this scenario, the black hole's event horizon would not be an ultimate barrier but a boundary separating its interior from what lies "outside"—our observable universe in its nascent stages.

This perspective reinterprets spaghettification, a well-known, terrifying black hole phenomenon. As an object nears a black hole, the differential gravitational pull stretches it vertically and compresses it horizontally, like spaghetti. Classically, this is destructive. However, within our hypothesis, this extreme stretching and deformation of spacetime and matter could be re-envisioned not as annihilation but as a transformative process. Under specific quantum gravitational conditions, this tidal stretching might be the mechanism that disperses and expands matter, facilitating its emergence as the building blocks of a new, outward-expanding cosmic structure.

This hypothesis aligns with other speculative cosmological models, particularly those involving cyclic or bouncing universes. In such a model, the cosmos undergoes endless cycles of expansion and contraction, where a Big Crunch transitions into a Big Bounce. Our hypothesis proposes a specific mechanism for this bounce: the collapse of matter into a black hole, leading not to a final end, but to the birth of a new universe from its interior. The infall of matter into a black hole could represent the contracting phase of a cosmic cycle, with the subsequent expansion from its singularity initiating a new universe. This suggests a potentially nested or cyclical cosmic history, where universes are born from the remnants of previous cosmic epochs, possibly even from within the remnants of collapsed stars or galaxies in an older, parent universe.

The theoretical frameworks supporting this audacious idea are diverse and at the forefront of physics research. Concepts from quantum gravity theories like String Theory, which posits that fundamental particles are vibrating strings, or Loop Quantum Gravity, which quantifies spacetime itself, offer potential mathematical tools and conceptual frameworks. These theories could describe the physics at the extreme densities and energies involved, potentially allowing a black hole singularity to transition into an expanding universe rather than a point of infinite collapse. While still under development and facing challenges in observational verification, these theories provide the foundational physics to explore the behavior of matter and spacetime under conditions far beyond terrestrial laboratory capabilities, pushing the boundaries of our understanding of the universe at its most fundamental levels.

The idea of our universe originating within a black hole is not merely fanciful speculation; it's a hypothesis that seeks to address perplexing enigmas in modern cosmology with a novel perspective rooted in established physical principles. This thought experiment, while speculative, compels us to reconsider the limits of our cosmic understanding and the potential for extraordinary origins. Subsequent sections of this chapter will delve deeper into this proposition, examining the intricate connections between the event horizon and the Big Bang, reinterpreting phenomena like spaghettification as potential cosmic rebirths, and exploring intriguing parallels with cyclic universe models. By engaging with these ideas, we aim to illuminate the intellectual landscape from which this hypothesis springs and to underscore its potential to reshape our comprehension of existence.

To fully appreciate the radical nature of the "universe within a black hole" hypothesis, it's essential to understand the foundational pillars of our current understanding of gravity and spacetime: Einstein's General Relativity. Published in 1915, this monumental theory revolutionized our perception of the universe by proposing a profound new way to understand gravity. Instead of Newton's concept of gravity as a force acting at a distance, Einstein revealed it as a geometric phenomenon, an intrinsic property of spacetime itself. He described spacetime not as a static backdrop, but as a dynamic, four-dimensional fabric—a pliable sheet interwoven with the dimensions of space and time. This fabric is constantly shaped, molded, and warped by the presence of mass and energy.

To visualize this intricate relationship, consider a stretched, taut, flat rubber sheet representing spacetime without significant mass or energy. If a heavy bowling ball is placed at the center, its mass creates a significant indentation or curvature. If a smaller marble is then rolled across the sheet, its path will curve inwards, spiraling towards the bowling ball, perhaps even settling into a circular orbit. This illustrates how Einstein's General Relativity describes gravity: massive objects like stars and planets warp the spacetime around them, and other objects, including light, follow these curves, perceiving the effect as the force of gravity. The more massive the object, the greater the curvature it induces, and the stronger the gravitational effect.

Mass and energy are fundamental principles that dictate the curvature of spacetime. This concept forms the foundation for understanding extreme gravitational phenomena, such as black holes. The more concentrated the mass in a given area, the more significant the warping of spacetime. This curvature determines the paths, or geodesics, that all objects and phenomena, including light, will follow. Geodesics are essentially the "straightest possible" paths through curved spacetime. In flat spacetime, a geodesic is a straight line, but in the curved spacetime around a massive object, geodesics are curved. This explains why planets orbit stars—they are following the curved paths in spacetime created by the star's mass, rather than being pulled by an invisible force. Similarly, light passing near a massive object, like a galaxy or a star, will bend its trajectory, a phenomenon known as gravitational lensing. These curved paths are precisely what matter and energy would follow if they were to fall into a black hole.

Einstein's theory is a rigorous mathematical framework, built upon complex equations that precisely relate the distribution of mass and energy to the curvature of spacetime. While a deep dive into the tensor calculus and differential geometry underpinning General Relativity is beyond the scope of this introductory chapter, it is crucial to acknowledge that these mathematical underpinnings are the bedrock upon which these concepts are built. The predictive power of General Relativity has been astonishingly successful, accurately describing phenomena from the subtle anomalies in Mercury's orbit to the bending of starlight during solar eclipses.

Central to Einstein's formulation of gravity is the Equivalence Principle, which states that the effects of gravity are indistinguishable from the effects of acceleration. For instance, if you are in a closed elevator accelerating upwards in space, you would feel a force pushing you down, similar to gravity. Conversely, the sensation in a stationary elevator on Earth experiencing gravity would be the same as in an accelerating elevator in deep space. This seemingly simple principle has profound implications, suggesting that gravity is not a fundamental force like electromagnetism, but rather a manifestation of spacetime geometry. It bridges the gap between the gravitational field and spacetime curvature, providing a crucial piece for understanding object behavior in strong gravitational fields.

To solidify the theoretical framework of General Relativity and its relevance to our cosmic hypotheses, it is imperative to touch upon the observational evidence that has repeatedly confirmed its validity. The anomalous precession of Mercury's orbit, a slight deviation from Newtonian gravity that puzzled astronomers for decades, was perfectly explained by Einstein's theory. During solar eclipses, astronomers observed that starlight passing near the Sun was indeed bent, precisely as predicted by the curvature of spacetime caused by the Sun's mass. More recently, the detection of gravitational waves—ripples in the fabric of spacetime caused by cataclysmic events like the merger of black holes—provided direct and unambiguous evidence for the dynamic nature of spacetime and the predictions of General Relativity. These confirmations, among many others, lend immense credibility to Einstein's theory, making the extrapolation of its principles to extreme environments like black holes a scientifically grounded endeavor. It is this robust theoretical foundation and overwhelming observational support that allows us to explore the more speculative, yet scientifically motivated, possibility of our universe being contained within a black hole.

The anatomy of a black hole, with its unique and extreme characteristics, provides the conceptual framework for our central hypothesis. Understanding these features is crucial to grasping how a universe might emerge from such an object. The most common pathway to black hole formation, at least for stellar-mass black holes, is the dramatic end-of-life event for massive stars. When a star significantly more massive than our Sun exhausts its nuclear fuel, the outward pressure from fusion that counteracts the inward pull of gravity ceases. Without this support, the star's core collapses catastrophically under its own immense weight. This collapse often triggers a spectacular supernova explosion, blasting the star's outer layers into space. However, if the remnant core, after the supernova, is sufficiently massive—exceeding a critical threshold known as the Tolman-Oppenheimer-Volkoff limit—no known force can halt its gravitational collapse. It continues to shrink, compressing its mass into an ever-smaller volume.

Black holes are characterized by a central singularity and an event horizon. The singularity, a point of infinite density and zero volume as predicted by General Relativity, represents a breakdown of known physics where spacetime curvature is infinitely strong. All matter and energy crossing the event horizon are drawn to this point and crushed. While a mathematical boundary, the singularity is also a focus for theoretical exploration, as quantum effects are expected to resolve the classical infinities.

The event horizon is a conceptual boundary in spacetime—a point of no return. It is not a physical surface, but rather a one-way membrane that traps anything, including light, once crossed. The size of the event horizon is determined by the black hole's mass, described by the Schwarzschild radius where escape velocity equals the speed of light. This boundary is crucial to the hypothesis that our universe might have originated from such a point.

Black holes vary in size and origin. Stellar black holes, a few to tens of solar masses, form from collapsing massive stars. Supermassive black holes, millions to billions of solar masses, reside at galactic centers, including the Milky Way's. Their formation involves gas and dust accretion, and possibly mergers of smaller black holes. Hypothetical primordial black holes, formed in the early universe, could be immense, potentially accounting for the vastness of our cosmos.

Falling into a black hole offers theoretical insights into extreme physics. Approaching a stellar-mass black hole, immense tidal forces cause "spaghettification," stretching an object vertically and compressing it horizontally. An infalling observer would cross the event horizon without immediate notice, their fate sealed. A distant observer, however, would see the object slow, redden, and fade, never quite crossing due to time dilation. These extreme conditions—intense gravity, distorted spacetime, and the ultimate fate at the singularity—form the basis for exploring the possibility of a cosmic birth.

This understanding of General Relativity and black hole anatomy leads to the conventional narrative of the Big Bang. The standard cosmological model (Lambda-CDM) posits that the universe originated from an extremely hot, dense state approximately 13.8 billion years ago. The Big Bang was an expansion of space itself, not an explosion. From this initial singularity, the universe rapidly expanded and cooled, forming fundamental particles, atoms, and eventually galaxies, stars, and clusters. The cosmic microwave background (CMB) radiation, a uniform glow, is considered strong evidence, representing the afterglow of the Big Bang when the universe was about 380,000 years old.

The Big Bang model, while largely successful, faces several unresolved questions. The horizon problem asks how regions of the universe, too far apart to have ever interacted, exhibit remarkably similar temperatures, as observed in the Cosmic Microwave Background (CMB). Another puzzle is the flatness problem: the universe is incredibly flat, a state that would require an implausibly precise initial condition to avoid rapid collapse or expansion. Furthermore, the nature of dark matter and dark energy, which constitute about 95% of the universe's mass-energy, remains a profound mystery, inferred only through their gravitational effects.

These unanswered questions highlight the need for alternative cosmological theories. This book proposes an audacious idea: our observable universe originated from within a black hole. This hypothesis suggests that the Big Bang was the result of matter and energy being compressed to extreme densities within a black hole and subsequently being expelled or reformed into an expanding cosmos. This concept posits that the singularity at the heart of a black hole might not be an endpoint, but a transitional phase—a cosmic crucible from which new universes are forged. This offers a potential mechanism for the origin of space, matter, energy, and the fundamental laws of our cosmos, all stemming from the intense gravitational conditions within a black hole.

The evidence for an expanding universe began with Edwin Hubble's observations in the late 1920s. He noted that light from distant galaxies is redshifted, meaning it shifts towards longer, redder wavelengths. Hubble's Law correlated this redshift with galactic distance, showing that the farther a galaxy is, the faster it recedes from us. This observed recession is a key indicator of an expanding universe, where the space between galaxies stretches over time. This foundational discovery underpins the Big Bang theory, demonstrating that our universe is dynamic and evolving.

The Cosmic Microwave Background (CMB) radiation provides further evidence. It is a nearly uniform bath of photons at about 2.7 Kelvin. However, sensitive instruments have detected tiny temperature fluctuations within the CMB, on the order of parts per hundred thousand. These minute anisotropies are crucial, as they represent slight density variations in the early universe, acting as the "seeds" for cosmic structure. Over billions of years, these denser regions attracted more matter through gravity, eventually forming the galaxies and galaxy clusters we see today. The CMB thus serves as a powerful testament to the universe's initial hot, dense state, its subsequent expansion and cooling, and is considered a relic of creation itself.

In the moments and minutes following the Big Bang, the very early universe was a realm of unimaginable energy and rapid transformations. During the Planck epoch, all four fundamental forces—gravity, electromagnetism, and the strong and weak nuclear forces—are thought to have been unified. As the universe expanded and cooled, these forces separated, and fundamental particles like quarks and leptons emerged.

Within the first few minutes, Big Bang nucleosynthesis occurred, where protons and neutrons fused to form the nuclei of the lightest elements: primarily hydrogen and helium, with trace amounts of lithium. The observed abundance of these elements in the universe today aligns remarkably well with the predictions of the Big Bang model, providing strong evidence for this cosmic origin story.

This hypothesis posits a fundamental connection between a black hole's event horizon and the Big Bang's initial singularity. The core idea suggests that as matter and energy traverse the event horizon, the intense tidal forces and quantum effects could initiate a "bounce" or a new phase of expansion. Rather than destruction, this matter and energy might transform, generating a new, expanding universe. This perspective aims to bridge black hole physics and cosmic origins, suggesting continuity between these phenomena, where the event horizon acts as a dynamic interface.

This interpretation also offers a fresh perspective on spaghettification. Traditionally seen as the destructive stretching of matter near a black hole's singularity, this extreme deformation, under quantum gravity, could be re-envisioned as a generative process. The intense stretching and shearing might facilitate the dispersal and expansion of highly compressed matter, leading to a new, outward-expanding universe. This recontextualizes a known black hole phenomenon as a potential origin for cosmic structures and observed expansion.

The hypothesis aligns with and draws parallels from cyclic or bouncing universe models, which propose an endless cycle of expansion and contraction. Our hypothesis offers a specific mechanism for such a bounce: the collapse of matter into a black hole. In this view, infalling matter represents the contracting phase of a cosmic cycle, and the subsequent expansion from the black hole's singularity becomes the Big Bounce, igniting a new universe. This paints a picture of a potentially cyclical or nested cosmic history, where universes emerge from the remnants of previous cosmic epochs, possibly even from within black holes of a larger cosmic structure.

To ground this radical idea, it's essential to consider supporting theoretical physics concepts. These include speculative but active research areas like quantum gravity theories (e.g., String Theory and Loop Quantum Gravity) and models exploring matter and spacetime under extreme conditions. These theoretical endeavors seek to reconcile General Relativity with quantum mechanics, crucial for understanding physics at the heart of a black hole or the Big Bang's initial moments. While still developing and facing experimental verification challenges, these theories offer potential tools and frameworks that might allow a black hole singularity to transition into an expanding universe, rather than merely an endpoint of collapse. Thus, this unconventional hypothesis is rooted in the ongoing quest to understand the universe at its most fundamental levels, drawing from the frontiers of theoretical exploration.

The proposition that our universe originated within a black hole, though speculative, compels a re-examination of our cosmological understanding. This hypothesis attempts to integrate the extreme physics of black holes with the grand narrative of cosmic genesis, potentially explaining some of our deepest mysteries. Future exploration will focus on how this paradigm-shifting idea might influence interpretations of existing observational data, potentially shedding light on dark matter and dark energy, and offering solutions to persistent cosmological puzzles like the fine-tuning of fundamental constants.

We will investigate whether anomalies in the cosmic microwave background (CMB), the Big Bang's faint afterglow, could be explained by an origin within a black hole. The hypothesis suggests that specific conditions of matter and energy dispersal from an event horizon might imprint unique signatures onto the CMB radiation, offering an alternative interpretation of patterns that currently challenge the standard model. For instance, could specific anisotropies or polarization patterns in the CMB be the residual imprint of the violent, yet potentially structured, ejection of matter from a black hole singularity? This section aims to find explanatory power within existing observational data, exploring how this novel model might account for phenomena that current theories struggle to fully explain. It calls for re-examining familiar datasets through a new theoretical lens, searching for hints that our cosmic origins might be more intimately linked to black hole physics than previously imagined.

The persistent enigmas of dark matter and dark energy offer fertile ground for exploring this hypothesis. These elusive components, which constitute the majority of the universe's mass-energy, remain fundamentally mysterious. Could they be remnants or byproducts of the initial transition from a black hole's interior to our expanding universe? The hypothesis suggests that dark matter, with its gravitational influence but lack of electromagnetic interaction, might be linked to the exotic quantum states of matter within a black hole's singularity. Similarly, dark energy, driving the universe's accelerating expansion, could be an emergent property of spacetime itself, a result of the initial energetic expulsion from the black hole. Considering these possibilities opens new perspectives on the composition and behavior of these dominant yet invisible cosmic constituents, potentially connecting them to the proposed generative process.

The "fine-tuning" problem in cosmology, which highlights the precise calibration of fundamental constants for life's existence, is also relevant. Why do constants like the gravitational constant, cosmological constant, or elementary particle masses possess their exact values? Slight variations would render the universe vastly different, possibly incapable of supporting life. Could the conditions within a black hole progenitor naturally lead to such a finely tuned universe? The hypothesis suggests a mechanism where physical laws and constants are established during the transition from the black hole's interior to the expanding universe. This offers a potential explanation for this cosmic coincidence, implying that fine-tuning is not random but a consequence of the physics governing our cosmos's birth from a black hole. The seemingly impossible precision of these values, tailored for our existence, could indicate a specific genesis event.

Furthermore, we will investigate if observed anomalies in galaxy formation or distribution align with a black hole origin. While the standard model generally explains large-scale structures like the cosmic web, are there subtle patterns or correlations better explained by matter dispersal from a specific, confined source? The hypothesis suggests that matter distribution and initial structure formation could bear the imprint of specific dynamics of matter ejection from an event horizon. We will consider if certain galactic arrangements or evolutionary paths are better explained by this alternative model than by the standard cosmological narrative, seeking subtle yet significant observational hints.

Finally, gravitational waves present an exciting avenue for testing this hypothesis. The violent and energetic processes of a theoretical black hole singularity transitioning into an expanding universe could generate unique gravitational wave patterns. These primordial spacetime ripples might carry signatures distinct from those predicted by the standard Big Bang model. Future gravitational wave observatories, with their increasing sensitivity, could discern such primordial signals. Detecting a specific type of gravitational wave background, perhaps with unique frequency distributions or polarization, could directly support or refute the hypothesis, powerfully probing the universe's earliest moments and its potential black hole genesis.

The quest to understand our universe's origins often leads to the cosmos's most extreme and enigmatic objects: black holes. The hypothesis that our universe might be contained within a black hole is deeply intertwined with the need for a comprehensive theory of quantum gravity. This is because both the singularity within a black hole and the initial moments of the Big Bang represent regimes where both gravity and quantum mechanics are paramount. Our current, highly successful theories—General Relativity for gravity and Quantum Mechanics for the very small—are incompatible in these extreme environments. General Relativity breaks down at the singularity, predicting infinite densities and curvatures, while Quantum Mechanics, governing particle behavior, does not inherently incorporate gravity.

A unifying theory of quantum gravity, capable of merging these two pillars of modern physics, is critical for understanding the physics at the heart of a black hole and the universe's initial moments. Such a theory is essential for a consistent mathematical framework describing the proposed transformation of a black hole's interior into an expanding universe. Without a unified theory, our understanding of these extreme conditions remains incomplete, relying on extrapolations and educated guesses. Theories like String Theory, suggesting fundamental particles are tiny vibrating strings, or Loop Quantum Gravity, quantifying spacetime itself, are ongoing attempts to bridge this gap, offering insights into how spacetime might behave at the Planck scale, where the smooth fabric of General Relativity might give way to a quantized, granular structure.

The event horizon of a black hole is another area where quantum mechanics is predicted to play a crucial role, with these quantum effects potentially directly relevant to our hypothesis. One of the most famous predictions from the intersection of quantum mechanics and General Relativity in black holes is Hawking radiation. Stephen Hawking proposed that black holes emit faint thermal radiation due to quantum fluctuations near the event horizon. These fluctuations involve spontaneous creation and annihilation of particle-antiparticle pairs. Occasionally, one particle falls into the black hole while the other escapes, carrying away energy. This process, while incredibly slow for astrophysical black holes, suggests the event horizon is a dynamic region where quantum effects are significant. This discussion will consider how these quantum effects, leading to Hawking radiation, might also contribute to the proposed transformation of matter and energy within a black hole into an expanding universe.

The singularity problem—the point of infinite density predicted by classical General Relativity at a black hole's center—is a major hurdle that quantum gravity theories aim to resolve. Instead of an infinitely dense point, these theories suggest the singularity might be replaced by a region of extremely high, but finite, density and curvature, where quantum effects dominate. This could lead to a "bounce," where infalling matter, instead of being crushed into oblivion, is repelled or reorganized in a way that leads to expansion. If such a bounce can occur, it opens the possibility for a black hole to serve as a progenitor for a new universe, with the event horizon acting as a boundary separating the internal cosmic structure from the external universe.

The information loss paradox highlights a fundamental conflict between quantum mechanics, which dictates that information cannot be destroyed, and classical predictions that suggest anything falling into a black hole is permanently lost. Resolving this paradox is crucial for our hypothesis, particularly if information from a "parent" universe is transferred during the black hole genesis process. Concepts like black hole complementarity and the holographic principle offer potential solutions, suggesting that information might be encoded in Hawking radiation, preserved within the black hole, or even represented on its boundary. These ideas are highly speculative but provide intriguing possibilities for how a universe's complexity and information could be contained within or emerge from a black hole's dynamics.

Developing a robust mathematical framework for a universe originating from a black hole is a monumental challenge. It requires integrating General Relativity's description of large-scale spacetime with quantum mechanics' principles governing the smallest scales. The primary hurdle lies in describing the physics at the black hole singularity, where current theories break down due to infinite spacetime curvature. A successful model would likely incorporate a quantum gravity theory that replaces the singularity with a region of finite, extreme density and quantum fluctuations, allowing for a transition rather than an endpoint. This necessitates new mathematical formalisms to model the birth of spacetime and matter from such extreme conditions.

To explore these complex theoretical scenarios, scientists employ sophisticated computational simulations. These simulations aim to replicate the extreme conditions near a black hole singularity and the subsequent expansion characteristic of a universe's birth. By defining initial parameters based on theoretical models (e.g., density, temperature, quantum properties under immense pressure) and evolving these conditions through complex algorithms, researchers can visualize spacetime warping, matter dispersal, and the emergent properties of the simulated cosmic environment. This allows for insights into the feasibility and consequences of a black hole-to-universe transition.

Crucially, any scientific hypothesis must be validated against observational data. For the black hole universe hypothesis, this means comparing the results of mathematical models and simulations with what we observe in our actual universe. Does the simulated expansion history match the observed Hubble expansion and age of the universe? Do the simulated early-universe signatures, like Cosmic Microwave Background (CMB) patterns or light element abundances, align with data from telescopes and satellites? This rigorous process of validation and falsification is the bedrock of scientific progress. Discrepancies necessitate model revision or abandonment, while alignment provides strong support.

The black hole universe hypothesis is one of several alternative cosmological models addressing shortcomings in the standard Big Bang model. Comparing its strengths and weaknesses with other frontiers, such as multiverse theories, cyclic universes, or modified gravity theories, helps clarify its unique contributions and challenges. For instance, while the Big Bang model explains much, it struggles with the horizon and flatness problems. The black hole hypothesis offers its own potential solutions, suggesting that the observed uniformity and flatness might be inherent properties of matter ejected from its black hole progenitor. This comparative analysis contextualizes the hypothesis within broader cosmological research, highlighting its novel insights and areas of competition.

Looking forward, advancements in quantum gravity will be paramount in refining and validating mathematical models for a black hole universe. The quest for a unified theory of everything, integrating quantum mechanics and general relativity, is central. Breakthroughs in understanding Planck scale spacetime, matter behavior under extreme gravity, and singularity properties could provide the theoretical framework needed to accurately describe a black hole origin for our universe. The dynamic interplay between theoretical progress and continuous observational data from next-generation telescopes will drive progress, potentially confirming or refuting this radical hypothesis and expanding our cosmic understanding.

Beyond its scientific implications, the hypothesis that our universe resides within a black hole carries profound philosophical and existential weight. If true, it drastically alters our perception of cosmic scale and belonging. Are we merely a component of a larger, perhaps incomprehensible, structure? Does this confinement diminish or enhance our universe's significance? This concept challenges anthropocentric views, suggesting our cosmic home is one among many, nested within a grander cosmic architecture. It redefines our cosmic address, prompting questions about uniqueness and existence.

This leads us to ponder the fundamental nature of reality. What does it mean for our universe to be "inside" something else? Could an "outside" exist, and what would that entail? This hypothesis opens the door to contemplating nested realities—layers of existence imperceptible to us. Such a structure has vast philosophical implications, potentially suggesting multiple levels of reality, each with its own physical laws or emergent properties. This invites deep contemplation on the fundamental structure of existence and the inherent limitations of our perception and knowledge, forcing us to consider that our perceived reality might be a mere segment of a grander, stranger cosmic arrangement.

The question of our universe's origin within a black hole also touches upon debates concerning determinism, free will, and cosmic causality. If our universe's existence and fundamental laws are intrinsically tied to processes within a black hole, does this imply a deterministic unfolding of cosmic events? Could the causal chains governing our existence extend beyond our observable universe, influencing our reality's conditions and constants? Exploring these connections prompts us to consider how a "parent" cosmic entity's past might intricately shape our universe's present and future, creating a complex, incomprehensible web of cause and effect. The very concepts of agency and free will might need re-examination in light of such vast, interconnected cosmic causality.

The profound concept of our universe existing within a black hole prompts a wide range of philosophical responses, influencing our continuous search for meaning. This idea can inspire awe at the cosmos' complexity, fostering humility and a sense of cosmic interconnectedness, suggesting our existence is part of a larger, dynamic cycle. Meaning, therefore, might be found in our participation within this vast, nested cosmic system, rather than in isolation.

This chapter ultimately celebrates the scientific curiosity that drives such radical hypotheses, emphasizing the importance of pushing the boundaries of our understanding despite challenges and skepticism. The pursuit of knowledge, even concerning outlandish ideas, is a fundamental aspect of both science and the human experience, propelling discovery and constantly redefining the limits of our understanding. The exploration of the black hole universe hypothesis, regardless of its eventual validation, exemplifies this ongoing quest to unravel the universe's deepest enigmas and reshape our perception of reality.

The journey through the black hole universe hypothesis concludes with a forward-looking perspective on cosmology's future, focusing on the scientific endeavors required for rigorous testing. The scientific method is a dynamic process, and while definitive evidence for this hypothesis remains elusive, observational strategies and theoretical developments offer a path forward. Key avenues of investigation include gravitational wave astronomy, which probes violent cosmic events, and ultra-precise measurements of the Cosmic Microwave Background (CMB), offering insights into the universe's infancy. The discussion highlights the specific data needed, from subtle CMB anomalies to distinct gravitational wave patterns, to support or reject the black hole origin of our cosmos, thus guiding future research.

The pursuit of definitive evidence is deeply linked to advancements in observational technologies. Next-generation ground-based and space-borne telescopes, with their unprecedented sensitivity and resolution, will be crucial for detecting faint signals, probing deeper into cosmic history, and discerning subtle features that may hold clues to our origins. Similarly, upgrades to gravitational wave detectors like LIGO, Virgo, and the planned LISA mission will expand our ability to observe a wider range of cosmic events and potentially detect primordial gravitational wave signatures predicted by theoretical models of a black hole genesis. These powerful new tools will enable a more direct and precise exploration of the universe's earliest moments.

Crucially, progress in understanding quantum gravity is paramount for validating or discarding the black hole universe model. The mathematical models describing the hypothetical transition from a black hole singularity to an expanding universe heavily rely on quantum gravity theories. As theoretical physicists refine frameworks like String Theory and Loop Quantum Gravity, and explore new avenues such as causal dynamical triangulations or asymptotically safe gravity, they provide the necessary theoretical foundation. Breakthroughs in understanding spacetime at the Planck scale, matter behavior under extreme gravitational conditions, and the fundamental properties of singularities could offer the precise mathematical language to accurately describe the universe's origin if it indeed occurred within a black hole. The ongoing quest for a theory of everything is central here, promising to resolve the incompatibilities between quantum mechanics and general relativity that currently limit our understanding of these extreme scenarios.

Beyond the specific focus on the black hole universe hypothesis, this section broadens the perspective to encompass other exciting frontiers in cosmology, showcasing the vibrant and dynamic nature of modern scientific inquiry. Research into the multiverse, dark energy, and potential modifications to the laws of gravity are active areas of research that could shed light on cosmic evolution. By examining these diverse avenues of inquiry, we gain a deeper appreciation for cosmology's multifaceted approach in its quest for a complete understanding of the cosmos. Each frontier presents unique challenges and potential breakthroughs, contributing to a richer and more comprehensive picture of reality.

The universe, a vast expanse of galaxies, nebulae, and the void, has always fueled human curiosity. From ancient stargazers to modern instruments, our quest to understand our cosmic home defines our civilization. We've charted constellations, understood stellar life cycles, and theorized about the universe's birth, evolution, and fate. Yet, profound mysteries remain, hinting at phenomena beyond our current grasp. This leads us to a radical concept: the possibility that our observable universe exists within a black hole.

This idea isn't presented as fact but as a compelling theoretical frontier, born from the intersection of established physics and speculative thought. It invites us to question our cosmic origins and the nature of the seemingly boundless space we inhabit. The goal is to spark curiosity, encouraging an open-minded exploration of our familiar universe as a unique, internal realm, sculpted by forces we are only beginning to comprehend.

To appreciate this hypothesis, we must first understand our current, widely accepted cosmological model. We've journeyed from the Big Bang's initial moments of extreme density and temperature, through rapid expansion and cooling, to the formation of subatomic particles, atoms, stars, and galaxies. The cosmic microwave background radiation (CMB) provides strong evidence for a universe that began explosively and has been expanding and cooling ever since. This model, built on decades of observation and refinement, serves as the foundation upon which we can explore the unconventional idea of a universe within a black hole. Establishing this baseline is crucial to grasping the radical nature of the proposition.

In this context, black holes, traditionally seen as cosmic anomalies or ultimate graveyards, begin to take on a more complex and potentially generative role. We've perceived them as gravitational traps, crushing matter and light into an infinitely dense singularity hidden behind an event horizon. However, deeper theoretical understanding reveals a more nuanced picture. The intense gravity, enigmatic singularity, and definitive event horizon are not just features of destruction; they are also characteristics that make black holes potent subjects for radical cosmological theories. The aim is to shift perspective, considering the possibility that these phenomena, far from being cosmic vacuum cleaners, might be cradles of new realities—the very wombs from which entire universes could be born.

The genesis of such an outlandish idea—our universe as an interior cosmos within a black hole—emerges from the fringes of our current theoretical understanding. It arises not from a void of scientific reasoning but from unexplored avenues and persistent inconsistencies within existing cosmological models. Puzzles like the Big Bang singularity's origin or the nature of spacetime at extreme densities naturally lead inquisitive minds to consider unorthodox explanations. This hypothesis is not an isolated spark of imagination but the culmination of an intellectual lineage that consistently pushes scientific boundaries. Throughout history, radical cosmological theories have emerged when existing paradigms encounter seemingly insurmountable obstacles, sometimes paving the way for revolutionary breakthroughs. By outlining this intellectual journey, we highlight the scientific courage and curiosity that drives such exploration, tracing the idea's evolution from its roots in established physics to its current form as a compelling theoretical frontier.

The human quest to understand the universe, epitomized by exploring hypotheses like the black hole universe, reflects an enduring drive for knowledge. Scientific advancement isn't solely about definitive answers but also about the continuous cycle of questioning, hypothesizing, testing, and refining our understanding. This process perpetually reshapes our perception of reality and expands our intellectual horizons, revealing countless wonders yet to be discovered in the universe.

Exploring unconventional hypotheses, such as the black hole universe, holds immense significance, however fantastical they may initially seem. These explorations act as intellectual crucibles, rigorously testing current theories and exposing their limitations. Specifically, this hypothesis offers potential novel explanations for persistent cosmological puzzles, potentially unifying disparate observations, providing a new perspective on the Big Bang's enigmatic origin, and fundamentally altering our understanding of reality, causality, and existence. By venturing into these speculative realms, we expand scientific knowledge and deepen our appreciation for the universe's vastness and complexity.

Our current cosmological understanding is built upon Einstein's General Relativity and the Standard Model of particle physics, depicting a cosmos that began approximately 13.8 billion years ago with the Big Bang—the rapid expansion of spacetime. As the universe expanded and cooled, particles and then atoms formed, followed by the gravitational coalescence of matter into stars, galaxies, and the large-scale structures of the cosmic web observed today.

Key evidence supporting this narrative includes the Cosmic Microwave Background (CMB) radiation. This faint, pervasive glow is the afterglow of the Big Bang, dating to about 380,000 years post-Big Bang when neutral atoms formed. The CMB's remarkable uniformity, punctuated by tiny temperature fluctuations, provides crucial information about early universe density variations, which seeded cosmic structures. Mapping these anisotropies has been vital in refining cosmological models.

The expansion of the universe, famously evidenced by Edwin Hubble's observations in the late 1920s, is another cornerstone. Hubble noted that distant galaxies exhibit redshift, correlating directly with their distance: the farther a galaxy, the faster it recedes. Hubble's Law interprets this as direct evidence of spacetime stretching, implying a smaller, denser, hotter universe in the past, leading back to the Big Bang singularity.

Matter in the universe is arranged in an intricate cosmic web of immense galaxy filaments and clusters separated by vast, nearly empty voids. This structure resulted from gravity acting on initial small density fluctuations in the early universe, where slightly denser regions attracted more matter to form massive structures, while underdense regions expanded into voids. Understanding this large-scale structure's formation and evolution is a critical test for any cosmological model, including the black hole origin hypothesis.

Despite the standard Big Bang model's success, persistent questions and fine-tuning problems remain. The horizon problem asks how causally disconnected regions of the universe could have such similar CMB temperatures. The flatness problem questions why the universe appears spatially flat. Furthermore, the nature of dark matter and dark energy remains mysterious. These unresolved puzzles motivate exploring alternative cosmological scenarios, such as the radical hypothesis that our universe originated within a black hole, potentially offering a unified explanation for these deep cosmological enigmas.

The proposition that our universe originated within a black hole radically departs from the standard narrative, yet it is rooted in the physics of black holes and the early universe. It suggests the Big Bang singularity was not an absolute beginning but a transitional rebirth within a black hole's extreme confines. This perspective attempts to weave together gravity and quantum mechanics, offering a potential mechanism for the genesis of spacetime, matter, and the laws governing our cosmos.

A key element of this hypothesis reinterprets a black hole's event horizon not as an ultimate boundary, but as a separator of spacetime realms. In this model, matter and energy falling into a black hole transform at the singularity, emerging as a new, expanding universe. The event horizon thus demarcates the black hole's "interior" from the nascent, expanding cosmos, reframing black holes as potential cradles for new cosmic epochs rather than dead ends.

Phenomena like spaghettification, the stretching of objects due to extreme tidal forces, are re-examined. Instead of being purely destructive, these tidal forces, influenced by quantum gravity, might disperse and distribute matter in a way that seeds a new universe. The intense stretching and compression of spacetime could be the generative mechanism leading to the Big Bang's outward expansion, offering a fresh perspective on black hole physics' most dramatic theoretical consequences.

This concept resonates with cyclical or oscillating universe models, which propose endless cycles of expansion and contraction. In this black hole-centric view, matter collapsing into a black hole represents a contracting phase, and the subsequent emergence of a new universe from its singularity signifies the "bounce" initiating a new expansion. This suggests a potentially nested or recursive cosmic history, where universes are born from previous cosmic entities, perhaps even from black holes within a larger cosmic structure. This cyclical perspective addresses questions about the universe's ultimate origin, suggesting a continuous process rather than a singular beginning.

The theoretical underpinnings supporting such a hypothesis are at the forefront of physics research, particularly in the quest for a theory of quantum gravity. Theories like String Theory and Loop Quantum Gravity attempt to reconcile general relativity and quantum mechanics, providing a framework to understand matter and spacetime behavior under extreme conditions found at black hole singularities or during the Big Bang. These advancements are crucial for developing mathematical models that accurately describe the proposed transition from a black hole's interior to an expanding universe, offering a rigorous pathway to test this audacious idea. Thus, the development of these theories is intrinsically linked to the validation or refutation of the black hole universe hypothesis.

Essentially, this proposed framework aims to unify the universe's origin by linking it to the extreme physics of black holes. It posits that the Big Bang was not arbitrary but a consequence of specific physical processes within a black hole. By examining known black hole properties and Big Bang evidence, this hypothesis offers a provocative yet scientifically motivated avenue for exploring deep cosmic questions. It encourages us to look beyond traditional boundaries of understanding and consider that our entire observable reality might be intricately connected to the enigmatic nature of black holes.

The cosmic microwave background (CMB) radiation, a relic from when the universe was 380,000 years old, presents a remarkably uniform thermal bath of microwave photons permeating all of space. This faint afterglow of the Big Bang is compelling evidence for the standard cosmological model. Its near-perfect uniformity is punctuated by minute temperature fluctuations (parts per hundred thousand). These anisotropies are not random; they represent slight variations in the primordial universe's density of matter and energy. These tiny inhomogeneities, under gravity's relentless pull over billions of years, grew into the vast cosmic structures we observe today: galaxies, galaxy clusters, and the intricate cosmic web.

The intricate dance of cosmic structures, from vast filaments of galaxies to the voids between them, is a direct result of gravity acting upon the universe's initial density fluctuations. Regions that were slightly denser in the early cosmos exerted a stronger gravitational pull, drawing in surrounding matter. Over immense spans of time, these overdense areas expanded, accumulating more material, while the underdense regions stretched into the expansive, nearly empty voids we observe today. This hierarchical process, where smaller structures coalesce into larger ones, has sculpted the cosmic web. This cosmic tapestry, with its threads of galaxies and galaxy clusters and its expansive, relatively empty stretches, demonstrates that the distribution of galaxies is not random. Instead, it's a direct reflection of the universe's initial conditions and its subsequent gravitational evolution.

The universe's expansion, evidenced by the redshift of light from distant galaxies, indicates that spacetime itself is stretching. This expansion is not uniform across all scales. While gravity binds galaxies within a cluster, causing them to move together, the clusters themselves are generally moving apart due to the expansion of the intervening space. This cosmic expansion also influences structure formation, acting as a counterforce to gravity and slowing the growth of structures, especially in the early universe. Understanding the interaction between gravity and cosmic expansion is crucial for accurately modeling the formation and evolution of galaxies and the cosmic web.

The standard cosmological model, the Lambda-CDM (Lambda-Cold Dark Matter) model, proposes that the universe is predominantly composed of dark energy (Lambda) and cold dark matter, with ordinary baryonic matter (which forms stars, planets, and us) making up only a small fraction. Dark matter, which doesn't interact with light, is vital for structure formation, providing the gravitational framework for galaxies and clusters. Dark energy, conversely, is believed to be responsible for the accelerating expansion of the universe observed in recent cosmic history. The exact nature of both dark matter and dark energy remains one of the most significant unsolved mysteries in cosmology, driving much of current research.

The hypothesis that our universe originated within a black hole offers a novel perspective on these fundamental cosmological aspects. It suggests that the initial conditions leading to the observed structure and expansion of our universe could be intrinsically linked to the physics of black holes. For instance, the uniform temperature of the Cosmic Microwave Background (CMB) might be explained by the rapid and homogeneous dispersal of matter and energy from a black hole's singularity. Similarly, the specific distribution of matter that formed the cosmic web might bear the imprint of the dynamics of ejection from a black hole. This hypothesis also provides potential, though speculative, explanations for the nature of dark matter and dark energy, suggesting they might be remnants or emergent properties of the black hole genesis process. By offering alternative explanations for these key cosmological phenomena, this hypothesis challenges our current understanding and encourages consideration of novel cosmic origin stories. This ability to potentially explain existing observations in a new light makes the exploration of such radical ideas a vital part of scientific progress, pushing the boundaries of our knowledge and deepening our appreciation for the vast and complex nature of the universe.

The very foundation of our understanding of the universe is steeped in awe and mystery, with black holes at the core of this cosmic tapestry. For decades, these enigmatic entities have been portrayed as the ultimate cosmic destroyers, voracious maw-like regions in spacetime from which nothing, not even light, can escape. Our initial encounters with them in scientific discourse and popular imagination often depict inescapable gravitational prisons, cosmic graveyards where matter and energy meet their ultimate, irreversible end. We visualize them as points of no return, defined by an impenetrable boundary known as the event horizon. However, as we push the frontiers of theoretical physics and confront the extreme conditions that define these celestial behemoths, a more nuanced and, indeed, generative role begins to emerge. The very properties that make them so terrifyingly destructive also imbue them with a profound potential to act as cosmic cradles, sites of creation and transformation that could fundamentally alter our perception of cosmic origins.

To grasp this paradigm shift, we must first revisit the fundamental characteristics that define a black hole. At its heart lies the singularity, a point of infinite density and curvature where the known laws of physics break down. Within this infinitesimally small, yet infinitely dense region, all the mass of the black hole is believed to be concentrated. Encircling this singularity is the event horizon, a spherical boundary in spacetime. This is not a physical surface in the conventional sense, but rather a conceptual boundary marking the point of no return. Anything crossing this threshold, be it a stray photon or an entire star, is inexorably drawn towards the singularity. The immense gravitational pull that creates the event horizon is the defining feature of a black hole, warping spacetime so intensely that it effectively encloses a region from which escape is impossible. This extreme gravitational gradient leads to the phenomenon known as spaghettification, where an object approaching a black hole would be stretched and squeezed like spaghetti due to the differential gravitational forces acting on its nearer and farther parts.

These characteristics – the singularity and the event horizon, coupled with immense gravitational forces – are not merely descriptors of a cosmic abyss. Instead, they represent the very conditions that make black holes such compelling subjects for speculative cosmology. The breakdown of physics at the singularity, far from being an endpoint of scientific inquiry, is a signpost pointing towards new physics, likely a theory of quantum gravity that can reconcile the seemingly incompatible realms of general relativity and quantum mechanics. It is at these extreme scales, where gravity is paramount and quantum effects cannot be ignored, that radical new ideas about the universe's origins can take root. The event horizon, too, is subject to reinterpretation. If we consider the possibility that spacetime itself can exhibit quantum properties, then the event horizon might not be an absolute barrier, but rather a transitional boundary, a veil separating one realm of existence from another.

The immense gravitational forces at play within a black hole are a testament to the power of mass concentrated into an incredibly small volume. These forces are responsible for the warping of spacetime, bending light and distorting the paths of matter. While this warping is understood in terms of attraction and eventual engulfment, it is also possible to conceive of these forces, under the influence of quantum effects, as agents of transformation rather than mere destruction. Imagine, for a moment, the process of matter falling into a black hole. It is subjected to unimaginable tidal forces, stretched and compressed in a manner that effectively disperses its constituent particles. This dispersal, when viewed through the lens of quantum gravity, might not lead to annihilation, but rather to a re-organization and re-emission of energy and matter in a new form, perhaps seeding a nascent universe.

This re-imagining of black holes as cosmic wombs rather than cosmic tombs is not a sudden leap of faith, but a gradual evolution of thought driven by the persistent questions that our current cosmological models struggle to answer. The ultimate origin of the Big Bang singularity, the nature of spacetime at such extreme densities, and the very reason for the universe's expansion are all profound mysteries. The extreme physics of black holes, with their singularities and event horizons, offer a tantalizing potential to provide answers. The powerful gravitational forces could, in theory, generate the immense energy densities required for a Big Bang-like event. The singularity, rather than being an absolute beginning, could be a point of transition, a place where the compressed remnants of a previous cosmic entity are reformulated and ejected to begin a new cycle of expansion.

Consider the event horizon once more. If it is not an an absolute boundary, but rather a membrane separating different regions of spacetime, then what lies beyond it might not be a void, but a different kind of reality altogether. This aligns with the speculative idea that our universe could be the interior of a black hole that formed in a parent universe. In this scenario, the matter and energy that fell into the parent universe's black hole did not vanish but underwent a transformation, giving rise to our own expanding cosmos. The event horizon of that original black hole would then serve as the boundary of our observable universe, a frontier beyond which we cannot see, much like the event horizon of a typical black hole limits our view. This perspective elegantly connects the physics of black holes with the grand narrative of cosmic origins, transforming our perception of these objects from passive absorbers of matter into active progenitors of new realities. The universe, in this view, is not a singular, unrepeatable event, but part of a larger, perhaps cyclical, cosmic process where black holes play a pivotal role in the ongoing creation and evolution of existence.

The idea that our universe exists within a black hole, a cosmic entity typically seen as a destroyer of matter and energy, initially seems counterintuitive. It challenges our fundamental understanding of cosmology, which posits a singular Big Bang expanding into an empty void. However, this radical concept isn't mere speculation. It emerged from persistent theoretical puzzles and the drive to unify general relativity and quantum mechanics, showcasing the scientific spirit of inquiry that confronts the limits of current knowledge and follows theoretical inconsistencies to revolutionary conclusions.

This idea's origins lie in the inherent tensions within our most successful cosmic models. Cosmologists have long struggled with the Big Bang singularity. General relativity, our best theory for gravity and the universe's large-scale structure, describes the Big Bang as a point of infinite density and curvature, where current physics breaks down. This singularity, while a powerful classical description, poses a conceptual hurdle: what existed before it, or how did it arise? The very concept of "before" becomes problematic if time began at the Big

Bang. This lack of a clear origin for the singularity opened the door for alternative explanations.

Black hole singularities bear striking similarities to the Big Bang's cosmological singularity. Both are theorized points of infinite density and curvature where familiar laws of physics collapse. This analogy – the parallel between a black hole's core and our universe's origin – became a recurring theme. If a black hole represents a region where spacetime collapses into a singularity, could the universe's origin be a similar, albeit vastly larger, phenomenon? This wasn't about finding a universe-containing black hole, but about exploring the universal process of singularity creation to explain our cosmic genesis.

The history of cosmology is rich with seemingly insurmountable problems spurring radical reinterpretations. The Steady State theory, for example, though ultimately disproven, demonstrated the scientific community's willingness to entertain profound alternatives when existing paradigms faced challenges. Similarly, the early universe, particularly the moments after the Big Bang, pushed theoretical physics to its limits. The extreme energies and densities demanded a deeper understanding that would bridge general relativity's macroscopic world with quantum mechanics' microscopic world.

The quest for a theory of quantum gravity drives modern theoretical physics. General relativity, while excellent for large-scale gravity, fails at the Planck scale, where quantum effects are expected to dominate gravitational interactions. Black holes, with their singularities and event horizons, are extreme gravitational environments where both theories are relevant. Understanding the physics at and within a black hole singularity is a crucial test for any quantum gravity theory. It is in this theoretical crucible that concepts like the "universe within a black hole" could emerge as natural consequences of a more complete understanding of gravity and spacetime.

The concept of an "event horizon," the point of no return, possesses an intriguing duality. From an external perspective, matter falling into a black hole appears to slow and freeze at the event horizon, redshifted into oblivion. However, an observer falling through the event horizon experiences no immediate barrier, continuing their journey toward the singularity. This difference hinted that the event horizon might not be an absolute, impenetrable wall, but a dynamic boundary with more complex quantum mechanical properties. Could this boundary, in a quantum gravitational sense, veil distinct spacetime regions or even different universes?

Early explorations of these ideas were often speculative, drawing on general relativity's mathematical frameworks and nascent quantum field theory. Black hole thermodynamics, pioneered by Jacob Bekenstein and Stephen Hawking, revealed profound connections between gravity, thermodynamics, and quantum mechanics. The discovery that black holes possess entropy and emit Hawking radiation suggested they are dynamic, quantum mechanical systems. This quantum nature opened new avenues, shifting black holes from purely classical objects to entities whose quantum behavior might have far-reaching implications.

The idea that the singularity might be a point of rebirth, not annihilation, gained traction. If matter and energy aren't destroyed in a black hole but transformed and ejected elsewhere, where could they go? Within general relativity, the most elegant answer was another spacetime region. For a cosmological singularity, this "elsewhere" could be a different epoch of the same universe or, more radically, a different universe altogether. This conceptual leap required viewing spacetime not as a fixed stage but as a dynamic, interconnected entity capable of birthing new cosmic epochs or even entirely new cosmos.

The scientific courage to pursue such ideas was immense. For generations, black holes were ingrained in scientific and popular consciousness as cosmic drains, swallowing all into oblivion. Proposing they could be generative, the birthplace of new universes, was a bold departure. It challenged established intuitions and ventured into areas with scarce observational evidence and developing theoretical frameworks. Scientists like John Archibald Wheeler, who coined "black hole," fostered an environment where such audacious questions could be asked. Wheeler himself was fascinated by the "cosmic egg" concept, a previous universe's collapse birthing a new one, which deeply resonates with the "universe within a black hole" hypothesis.

The hypothesis developed through exploring various theoretical pathways. One path investigated wormholes, hypothetical spacetime tunnels connecting distant points or even different universes. While speculative, the mathematical possibility of traversable wormholes in certain solutions to Einstein's field equations, especially those involving exotic matter, provided a theoretical mechanism for an internal black hole singularity to connect to a different spacetime region. If the singularity was a gateway, matter falling into the black hole could potentially emerge in a new, expanding universe.

Another avenue considered spacetime's quantum nature. If spacetime is granular or foamy at the Planck scale, the event horizon might be a surface where intense quantum fluctuations facilitate information and matter transfer to another spacetime region. This aligns with the holographic principle, suggesting that a volume of space's information can be encoded on its boundary. For a black hole, its event horizon could be a holographic screen, where quantum gravitational processes at the singularity are projected or manifested as a new expanding universe.

The idea of a "universe within a black hole" isn't a sudden, radical concept, but rather a gradual evolution of scientific thought. This hypothesis arose from significant cosmological challenges such as the Big Bang singularity, the mathematical parallels between cosmological and black hole singularities, the quantum properties of black holes, and the theoretical exploration of concepts like wormholes and holography. Over decades, numerous physicists pursued these diverse lines of inquiry, which ultimately converged to paint a picture of the cosmos that is far more intricate and interconnected than previously imagined. Embracing this idea required both intellectual rigor and a willingness to accept that the universe's most destructive forces might also be its most creative. It represents a paradigm shift, prompting us to look beyond the perceived finality of the event horizon and consider the possibility of new beginnings hidden within the cosmic darkness.

The proposition that our universe could exist within a black hole, while seemingly speculative, holds profound significance for cosmological research. Seriously considering such a notion is not a departure from rigorous scientific inquiry, but a necessary engagement with the deepest and most persistent enigmas in our current understanding of the cosmos. The value of this hypothesis lies not in its immediate provability, but in its powerful ability to foster new theoretical frameworks, resolve long-standing paradoxes, and fundamentally re-evaluate our perception of reality. It is through the courageous exploration of such unconventional ideas that science often achieves its most significant breakthroughs, pushing the boundaries of our knowledge into uncharted, yet potentially illuminating, territories.

At its core, the significance of this hypothesis lies in its potential to offer a unifying explanation for a range of seemingly disconnected cosmological observations and theoretical challenges. For instance, the enduring mystery of the Big Bang singularity, where our current laws of physics fail, has long been a central focus of cosmic inquiry. If, as some theories suggest, a black hole singularity is a transitional phase rather than an endpoint, a gateway to another region of spacetime, then the origin of our universe could be understood as a natural consequence of such a process. This perspective moves away from the problematic idea of a universe spontaneously appearing from an incomprehensible singularity, instead offering a framework where our cosmos's beginning is intimately linked to the extreme, yet understood, physics of gravitational collapse. The persistent, tantalizing clue has been the similarity between the singularities predicted by general relativity for the Big Bang and those found at the heart of black holes. By suggesting that our universe originated within a black hole, we are not merely drawing an analogy; we are proposing a physical mechanism, rooted in the behavior of spacetime and matter under extreme conditions, that could explain our cosmic genesis. This offers a potential solution to the "what came before the Big Bang?" question, not by invoking an infinite regression of past universes, but by suggesting a cyclical or generative process where the end of one universe is the beginning of another.

Furthermore, this hypothesis provides a fresh perspective for re-examining some of the most perplexing observations in modern cosmology. Consider, for example, the nature of dark energy and the accelerated expansion of the universe. While the cosmological constant (Lambda) successfully describes this phenomenon within the standard Lambda-CDM model, its physical origin remains deeply mysterious. The energy density associated with the vacuum of space, as predicted by quantum field theory, is vastly larger – by many orders of magnitude – than the observed value of the cosmological constant, a discrepancy known as the "cosmological constant problem." Could the presence of an external, higher-dimensional structure, such as the "parent" universe in which our black hole-born cosmos resides, influence the fundamental constants and energy densities of our own universe? The hypothesis offers a potential way to address this fine-tuning problem. If our universe is born from the collapse of a massive object in another universe, the conditions of that parent universe, its physical laws, and its energy landscape might imprint themselves upon our nascent cosmos, setting the initial values for parameters like the cosmological constant. This could explain why our universe appears so finely tuned for the existence of life, a concept often referred to as the anthropic principle. Instead of our universe being a statistical fluke, it could be a product of specific conditions in its progenitor.

The significance of this hypothesis also extends to its potential to bridge the persistent gap between general relativity and quantum mechanics, the two pillars of modern physics that have, thus far, resisted complete unification. Black holes are precisely the kind of extreme environments where both theories are expected to play crucial roles, and where their incompatibilities become most apparent. The singularity at the heart of a black hole, where spacetime curvature becomes infinite according to general relativity, is a region that quantum effects must, in principle, resolve. Similarly, the event horizon, while a classical concept, is a region where quantum phenomena like Hawking radiation become relevant. If our universe emerged from a black hole singularity, understanding the quantum processes that govern such an event would be paramount. This hypothesis compels us to develop a more complete theory of quantum gravity, one that can describe not only the interiors of black holes but also the very origin of spacetime and matter. It forces us to ask: what are the quantum gravitational rules that govern the birth of a universe from a singularity? Are these rules different from those that govern the collapse of matter into a black hole? The very act of exploring these questions pushes the boundaries of theoretical physics, prompting the development of new mathematical tools and conceptual frameworks, such as string theory or loop quantum gravity, that aim to reconcile gravity with quantum mechanics.

The notion that our universe exists within a black hole profoundly alters our understanding of the cosmos. This concept suggests a vast, interconnected multiverse, not born from quantum fluctuations but from the remnants of previous cosmic eras or stellar collapses. This paints a picture of a cyclical or generative cosmic architecture, where intergalactic "emptiness" might be an illusion, a facet of a more intricate reality. The very definitions of "outside" and "boundary" become complex; if our universe is within a black hole, its event horizon becomes our cosmic boundary. This perspective could lead to a reevaluation of cosmological models, potentially offering insights into dark matter, suggesting its gravitational influence might stem from the "parent" universe or the containing black hole, a highly speculative but powerful explanatory hypothesis.

This hypothesis also holds significant potential for addressing the arrow of time and the perceived asymmetry between cosmic creation and destruction. Black holes, traditionally seen as ultimate destroyers, are reframed as "cosmic wombs" from which universes emerge. This cyclical view suggests a universe with profound resilience and a tendency towards regeneration, resonating with various philosophical and spiritual traditions within a scientific framework. Furthermore, the long-standing problem of information loss in black holes could find a resolution here; if a universe is born from a black hole, information is not lost but reconstituted in a new cosmic context, with profound implications for causality and information conservation.

Moreover, this hypothesis encourages a more imaginative approach to theoretical physics. When faced with phenomena like the universe's origin or dark energy, entertaining ideas that challenge ingrained assumptions is crucial. The "universe within a black hole" hypothesis is precisely such an idea, pushing physicists to explore mathematical consistency and physical plausibility. This intellectual rigor, applied to seemingly outlandish concepts, can lead to unexpected breakthroughs, much like the exploration of wormholes or black hole thermodynamics. It serves as a modern example of how pursuing improbable ideas can illuminate fundamental questions.

Perhaps the most profound aspect lies in its implications for the nature of reality itself. If our universe is one among an infinite cosmic "menagerie," our understanding of existence is radically transformed. We become participants in a vast, interconnected cosmic tapestry, finding solace in the idea that cosmic cycles of creation and rebirth continue even if our universe ends. This also raises deep philosophical questions about space and time; if our universe has a finite boundary within an infinite "container" universe, it challenges our intuitive notions of scale. It could also offer new avenues for understanding the cosmological constant problem, suggesting that our observed "vacuum energy" might be a consequence of the larger spacetime and the process of "bubble nucleation."

Ultimately, the significance of the "universe within a black hole" hypothesis lies in its power to inspire new research and stimulate innovative thinking. It reminds us that our current understanding, while successful, is incomplete. Fundamental questions remain, and exciting discoveries often lie at the frontiers of our knowledge. This hypothesis, by offering a compelling, albeit speculative, alternative to persistent cosmic puzzles, encourages physicists to explore uncharted territories with renewed curiosity and intellectual courage. It is a testament to scientific inquiry, a journey that, by embracing the unknown and daring to ask "what if?", continues to unveil the breathtaking complexity and wonder of the universe, both known and beyond our current grasp. The intellectual landscape it opens is vast, promising future paradigm-shifting discoveries.

CHAPTER 2

Einstein's Legacy: Gravity and Spacetime

Before Einstein, Isaac Newton's theory largely shaped our understanding of gravity. Newton saw gravity as an invisible, instantaneous force acting between any two objects with mass, pulling them across space. This force held us to the ground, kept the Moon orbiting Earth, and tethered planets to the Sun. Newton's theory could accurately predict an apple's fall and the planets' celestial movements. However, Newton himself was puzzled by "action at a distance"—how objects could influence each other without a physical connection or intermediary. These profound questions were beyond his revolutionary framework.

Einstein, however, wasn't content with just describing gravity; he sought to understand its underlying mechanisms. His curiosity led him to question space and time, concepts long considered absolute and immutable backdrops. Newton's laws implicitly assumed a fixed stage. Einstein's groundbreaking insight was to propose that space and time are not separate, static entities but are interwoven into a single, dynamic, four-dimensional continuum: spacetime. This "fabric of the cosmos" is not a passive stage but an active, malleable entity that can be warped, curved, and influenced by the matter and energy it contains.

To grasp this shift, imagine a stretched rubber sheet representing spacetime. Place a heavy bowling ball (a massive object like a star or planet) at its center. This creates an indentation, a curvature, in the sheet, analogous to spacetime distortion caused by mass and energy. If you roll a smaller marble across the sheet, it won't travel in a straight line but will be deflected by the bowling ball's curvature, appearing to be pulled towards it. This is Einstein's description of gravity: not a force pulling objects together, but the consequence of objects following curves in spacetime, which are themselves created by mass and energy. Greater mass creates a deeper indentation and a stronger apparent gravitational pull.

This revolutionary idea—that gravity is a manifestation of spacetime's geometry—was the cornerstone of Einstein's General Theory of Relativity, published in 1915. It elegantly resolved the mystery of Newton's "action at a distance." According to Einstein, gravity's influence is not instantaneous; it propagates at the speed of light, like ripples on a pond. When a massive object moves or changes, it sends out "gravitational waves"—ripples in spacetime's fabric—that travel outward, carrying information about its motion. This differs greatly from Newton's instantaneous action and has profound implications. For example, if the Sun vanished, we wouldn't immediately feel gravity's effects. It would take approximately eight minutes—the time for light and gravitational waves to travel from the Sun to Earth—for our planet to begin freefalling into the void.

The mathematical framework Einstein developed for this geometric interpretation of gravity is the Einstein Field Equations. These ten complex, non-linear differential equations relate spacetime's curvature (represented by the Einstein tensor) to the distribution of mass and energy within that spacetime (represented by the stress-energy tensor). Essentially, these equations describe how matter and energy tell spacetime how to curve, and how spacetime's curvature tells matter and energy how to move. They are the heart of General Relativity, a testament to Einstein's genius in capturing the subtle dance between the universe's constituents and the stage on which they interact.

General Relativity, Einstein's geometric theory of gravity, has profoundly impacted our understanding of the universe, with its implications consistently supported by observational evidence.

One of the earliest and most famous confirmations came during the 1919 solar eclipse. Newtonian physics predicted starlight would travel in a straight line, but Einstein's theory posited that the Sun's immense mass would warp spacetime, causing light from distant stars to bend as it passed nearby.

Expeditions led by Arthur Eddington observed stars near the Sun's limb in slightly different positions than expected, confirming Einstein's prediction with remarkable accuracy. This observation propelled Einstein to international fame and established General Relativity as the new standard for understanding gravity.

Another critical prediction is gravitational redshift. General Relativity explains that gravity warps spacetime and affects the passage of time, causing time to run slower in stronger gravitational fields. Consequently, light emitted from a strong gravitational field (like a star's surface) loses energy as it travels into a weaker field, manifesting as a shift towards longer, redder wavelengths. This effect has been precisely measured using atomic clocks at different altitudes on Earth, demonstrating the subtle slowing of time and the resulting gravitational redshift at lower elevations.

General Relativity also provides the theoretical framework for black holes, regions of spacetime where gravity is so intense that nothing, not even light, can escape. These arise directly from Einstein's field equations: when a massive star exhausts its fuel and collapses, it can form an infinitely dense singularity surrounded by an event horizon. Crossing this boundary is a one-way trip, as spacetime curvature is so extreme that all paths lead to the singularity. Once a purely theoretical concept, the existence of black holes is now overwhelmingly confirmed by astronomical observations, from galactic supermassive black holes to direct imaging of Messier 87's black hole shadow.

The theory has vast implications for cosmology, the study of the universe as a whole. The Einstein Field Equations suggest a dynamic, non-static cosmos, either expanding or contracting. This radical idea was confirmed by Edwin Hubble's observations in the late 1920s, which showed distant galaxies moving away from us at speeds proportional to their distance, providing compelling evidence for the universe's dynamic nature. This expansion is not merely an explosion into existing space but an expansion of space itself, carrying galaxies along.

Furthermore, General Relativity is the bedrock of the Big Bang model, our best description of the universe's origin and evolution. This model posits that the universe began in an extremely hot, dense state and has been expanding and cooling ever since, with the very fabric of spacetime undergoing this expansion. The theory explains the abundance of light elements, the cosmic microwave background radiation (the Big Bang's afterglow), and the universe's large-scale structure. Without General Relativity, our understanding of cosmic history would be fundamentally incomplete, as it allows us to trace the universe's journey from its earliest moments to its present state, providing a consistent and powerful narrative of cosmic evolution.

General Relativity's beauty lies not only in its predictive power but also in its conceptual elegance. It offers a unified, geometric description of gravity, replacing the notion of a mysterious force with a deeper understanding of the interplay between matter, energy, and spacetime's structure. It has opened new avenues of scientific inquiry, from gravitational waves and black holes to the universe's origins and ultimate fate. It remains one of the most successful and rigorously tested theories in science, a towering achievement that continues to shape our perception of the cosmos and our place within it. Einstein's General Relativity portrays the universe not as a static backdrop, but as a dynamic, ever-evolving entity whose geometry is sculpted by everything within it—a breathtaking cosmic dance orchestrated by the fundamental laws of physics.

Albert Einstein's theory of General Relativity posits that gravity is not a traditional force, but rather a geometric property of the universe, emerging from the dynamic interplay between mass, energy, and spacetime. To grasp this concept, we must move beyond the Newtonian idea of an invisible pull and understand gravity as a consequence of spacetime being molded by matter and energy.

Imagine spacetime as a four-dimensional canvas. Any object with mass or energy, from a subatomic particle to a galaxy, imprints distortions—curves and warps—onto this canvas. These distortions then dictate the motion of everything else.

Consider a stretched rubber sheet: placing a heavy sphere on it creates a depression. If a smaller sphere is then rolled across the sheet, its path will naturally curve towards the larger sphere, following the slope of the depression. This isn't due to a mysterious force, but because the sheet's geometry has been altered, guiding the smaller sphere's trajectory. General Relativity proposes that mass and energy do the same to the four-dimensional fabric of spacetime. Greater concentration of mass or energy leads to more pronounced curvature, which in turn dictates the movement of other objects, including light. They aren't being pulled but are following the "straightest possible path" (a geodesic) within a curved geometry.

This fundamental principle explains phenomena Newtonian gravity could only describe. For instance, planets orbit the Sun not because of a direct pull, but because the Sun's immense mass warps the surrounding spacetime, and planets follow these natural contours. Similarly, the bending of light around massive objects, confirmed during solar eclipses, is a direct result of light traveling along geodesics in curved spacetime.

The Einstein Field Equations, a set of complex tensor equations, mathematically encapsulate this description. They link the geometry of spacetime (described by the Einstein tensor) with its contents (quantified by the stress-energy tensor, representing mass, energy, momentum, and stress). In essence, matter and energy tell spacetime how to curve, and spacetime's curvature tells matter and energy how to move. This elegant and self-consistent framework enables General Relativity to make precise, observationally validated predictions about gravity. The beauty of these equations lies in their profound ability to unify gravity, spacetime, and the universe's fundamental constituents.

One dramatic implication of this mass-energy-spacetime curvature is our understanding of black holes. These enigmatic objects represent the extreme limit of spacetime curvature. When a sufficiently massive star collapses under its own gravity after exhausting its nuclear fuel, its remnant core can be crushed into an infinitely dense singularity. This singularity is surrounded by an event horizon—a boundary in spacetime where the curvature is so extreme that escape velocity exceeds the speed of light, making it a point of no return.

Within the event horizon, all paths, including those of light, are directed inevitably towards the central singularity. Analogous to the rubber sheet, imagine a depression so deep and steep that nothing falling in can ever climb out—that's an event horizon. For an object falling into a black hole, time itself is distorted. A distant observer would see the object slow down, its light redshifted and dim, appearing to freeze indefinitely at the horizon. However, for the object itself, the journey across the event horizon would be ordinary, though it would face immense tidal forces that stretch and compress it, a process known as "spaghettification."

The existence of black holes, predicted by Einstein's field equations, has been robustly confirmed through astronomical observations, including their gravitational influence on surrounding matter, X-ray emissions from infalling gas, and the direct imaging of Messier 87's black hole shadow by the Event Horizon Telescope. These observations not only confirm black holes but also directly validate the profound spacetime distortions predicted by Einstein's theory.

The relationship between mass, energy, and curvature also applies to the universe at large. When applied cosmically, the Einstein Field Equations reveal that the universe cannot be static; it must be expanding or contracting. This was a groundbreaking departure from the then-prevailing view of a fixed universe. Edwin Hubble's subsequent discovery that distant galaxies are receding proportionally to their distance provided crucial empirical support for General Relativity's dynamic universe. This expansion is not an explosion into existing empty space, but rather an expansion of spacetime itself, carrying galaxies with it.

In this cosmic expansion, spacetime's curvature is not uniform. The large-scale distribution of matter and energy determines the universe's overall geometry. Cosmological models, derived from the field equations with various assumptions about the universe's contents, predict different fates for the cosmos based on its overall curvature. For instance, a positively curved universe (like a sphere) would eventually collapse. A zero-curvature universe (like a flat plane) would expand forever at a decreasing rate. A negatively curved universe (like a saddle) would also expand forever, but at an accelerating rate. Current observations suggest the universe is remarkably flat on large scales, though the influence of dark energy complicates this, leading to an accelerating expansion.

Einstein's theory of General Relativity posits that the universe's energy content, not merely its mass, dictates the curvature of spacetime. This concept stems from his famous equation, $E=mc^2$, which establishes the interchangeability of mass and energy. Consequently, all forms of energy contribute to spacetime's curvature, profoundly impacting phenomena like cosmic expansion. For example, the early universe's radiation energy density was crucial to its evolution and the formation of large-scale structures. Dark energy, a mysterious force accelerating the universe's expansion, further illustrates energy's fundamental role in shaping spacetime.

General Relativity employs complex mathematical tools like tensors, differential geometry, and curved manifolds to describe this curvature. While a detailed exploration of these is beyond the current scope, their power in precisely quantifying spacetime distortion caused by mass and energy is undeniable. These constructs enable physicists to accurately calculate object trajectories, light behavior, and universal evolution. The field equations concisely express the intricate relationship between the universe's physical properties (mass, energy, momentum) and its geometric attributes (curvature).

The theory's predictive accuracy is remarkable. Beyond known phenomena like starlight bending and gravitational redshift, General Relativity predicted gravitational waves –

spacetime ripples generated by accelerating massive objects, such as merging black holes or neutron stars. The direct detection of these waves by observatories like LIGO and Virgo represents a monumental achievement, providing a new lens into the universe. These waves carry vital information about the cosmos's most violent events, traversing vast distances unimpeded and offering an unprecedented way to observe otherwise invisible phenomena. Their detection, precisely matching theoretical predictions for frequencies and amplitudes, underscores Einstein's profound understanding of gravity.

At its core, Einstein's revolutionary theory hinges on the relationship between mass, energy, and spacetime curvature. This transforms gravity from a mysterious force into a geometric consequence of the universe's very structure. This geometric perspective is not just an abstract idea; it's a powerful predictive tool that has illuminated our understanding of extreme cosmic phenomena, from black holes' intense gravity to the vast sweep of cosmic evolution. It celebrates human ingenuity and our ongoing quest to comprehend the universe's fundamental workings. From this perspective, the universe is a dynamic, interconnected entity where matter and energy directly sculpt the stage for all cosmic events.

In General Relativity, the universe's stage is not static but a dynamic, four-dimensional tapestry called spacetime. Upon this ever-shifting tapestry, celestial bodies and light perform their cosmic ballet. The choreography is governed by Einstein's elegant geometric interpretation of gravity: objects follow the "straightest possible paths" or geodesics through this curved spacetime, rather than being moved by an invisible force.

To illustrate, consider the analogy of a stretched rubber sheet. Placing a bowling ball in the center creates a significant dip. If you roll a marble across the sheet, slightly off-center from the bowling ball, it will curve and spiral inwards towards the ball. From a two-dimensional view on the sheet, this appears to be the marble's most direct route given the surface's deformation. The bowling ball isn't actively "pulling" the marble; rather, the sheet's geometry dictates the marble's trajectory. The marble follows the "straightest" available path on a curved surface.

This precisely mirrors how General Relativity describes gravity. Massive objects like stars and planets warp the surrounding spacetime fabric, making it "non-flat." Any object or even a beam of light traveling through this curved region will endeavor to follow a geodesic – a path that locally appears straight in a curved geometry. Unlike simple straight lines in flat space, these "straight" paths are not perceived as such in our everyday experience.

Take Earth's orbit around the Sun. Newtonian physics explains this as the Sun's gravitational pull keeping Earth in constant acceleration. Einstein, however, offers a different perspective: the Sun's immense mass creates a deep curvature in surrounding spacetime. Earth simply follows the geodesics within this curved spacetime. It's not being pulled by an unseen force; it's following the most natural, straightest path available in the geometry the Sun has created. If spacetime were flat, Earth would move in a straight line indefinitely. But because spacetime is curved, that "straight" line becomes an elliptical orbit. Earth is, in essence, falling towards the Sun while simultaneously moving sideways at a velocity that continuously causes it to "miss" the Sun, thus maintaining its orbit.

Understanding Gravity Through Spacetime Geodesics

The concept of geodesics in curved spacetime is fundamental to understanding gravity, explaining phenomena from falling objects to the paths of light and planets. This principle asserts that objects, regardless of their mass, follow similar paths in the same gravitational field.

Experimental Verification: From the Moon to Light Bending

A classic demonstration of this principle is the simultaneous fall of a feather and a hammer on the Moon. Without air resistance, both objects hit the surface at the exact same time because their motion is governed by the Moon's spacetime curvature, not by their individual masses. Unlike Newtonian gravity, where gravitational force depends on mass (F=ma), General Relativity posits that acceleration is a consequence of spacetime's geometry.

The bending of light around massive objects further confirms this. Light, though massless, follows spacetime geodesics. When starlight passes near the Sun, the Sun's mass curves spacetime, causing the light to deviate from a straight line. This makes the star appear in a slightly different position, a phenomenon famously observed and confirmed during solar eclipses in the early 20th century. This observation was a critical validation of General Relativity, showing that light rays were not "pulled" by the Sun but were tracing the straightest possible paths through the curved spacetime.

The Mathematical Definition and Its Implications

Mathematically, a geodesic in curved spacetime is defined as a "straight" curve where its tangent vector remains parallel to itself when transported along the curve. In flat spacetime, this describes a straight line with constant velocity. However, in curved spacetime, the geodesic equation accounts for the curvature, meaning objects following a geodesic will experience acceleration even without traditional forces, because the "straightness" of their path constantly changes due to geometry.

These paths are not arbitrary; they represent the paths of least "proper time" between two spacetime points when influenced only by gravity, aligning with the principle of least action. This "time" is a complex concept reflecting the interwoven nature of space and time.

A Geometric View of the Universe

This geometric interpretation of gravity is both economical and powerful. Instead of an instantaneous force acting at a distance, General Relativity describes gravity as an intrinsic property of spacetime, caused by the presence of mass and energy. This elegantly explains why gravity affects all objects equally, including light. The universe becomes a dynamic entity where matter and energy sculpt the spacetime in which they move, and in turn, the shape of spacetime dictates the motion of objects.

Geodesics in Extreme Environments and Planetary Orbits

The implications of geodesics extend to extreme cosmic environments like black holes. Around a black hole, spacetime is so intensely curved that geodesics are severely distorted. The event horizon is where all outgoing geodesics are directed inwards, towards the singularity. Once crossed, an object is irrevocably committed to the black hole's center, not due to an infinitely strong pull, but because all "straight" paths within the event horizon lead inward.

Furthermore, geodesics explain the subtle deviations in planetary orbits, particularly Mercury's perihelion precession. Newtonian gravity couldn't fully account for this slight shift in Mercury's closest point to the Sun. General Relativity, by considering Mercury's motion along geodesics in spacetime curved by the Sun, precisely predicts this anomaly. Mercury's orbit is not a perfect ellipse in the Newtonian sense; it's a geodesic in curved spacetime, providing direct confirmation of Einstein's theory.

The universe operates on principles mirroring those found on a smaller scale. The distribution of matter and energy across the cosmos dictates the overall shape of spacetime, determining whether it is flat, positively curved (like a sphere), or negatively curved (like a saddle), depending on its total mass-energy content. The movement of galaxies and the universe's expansion are essentially the result of these large-scale geodesics—paths of least resistance. The cosmic microwave background radiation, a remnant of the early universe, also provides insights into spacetime geometry, with its observed patterns aligning perfectly with General Relativity's predictions.

Ultimately, geodesics are how the universe achieves "straightness" within a potentially curved framework. These natural trajectories, determined by spacetime's geometry, represent the most effortless paths. This fundamental concept transforms our understanding of gravity from an external force to an intrinsic property of the cosmos, highlighting the deep connection between mass, energy, and the fabric of reality. By comprehending these curved, yet inherently "straight," paths, we begin to unlock the universe's deepest mysteries.

A cornerstone of Einstein's General Relativity is the profound realization that gravity is not a mysterious force but rather a consequence of reality's fabric being warped by mass and energy. Building on this geometric view of gravity, we encounter the Equivalence Principle—a concept so elegantly simple yet incredibly significant that it forms the bedrock of the entire theory. This principle, born from Einstein's thought experiments, asserts that the effects of gravity are completely indistinguishable from the effects of acceleration.

To fully grasp the power of the Equivalence Principle, consider one of Einstein's famous thought experiments. Imagine being inside a sealed, windowless elevator at rest on Earth's surface, experiencing Earth's gravitational pull. If you release a ball, it falls at approximately 9.8 meters per second squared, and jumping results in the same downward acceleration. Everything feels normal, governed by the familiar force of gravity.

Now, imagine the same sealed elevator, with you and the ball inside, adrift in the vastness of intergalactic space, free from significant gravitational influence. Suddenly, the elevator accelerates upward at a constant rate of 9.8 meters per second squared, propelled by a powerful, invisible rocket engine. From your perspective within the elevator, the experience would be remarkably similar. Releasing the ball would make it appear to fall to the floor at 9.8 m/s^2, and jumping would push you back towards the floor with the same acceleration, simulating the effects of gravity. In theory, experiments performed within this accelerating elevator would yield identical results to those conducted in a stationary elevator on Earth.

This is the core of the Equivalence Principle: no experiment performed within the sealed confines of the elevator can distinguish between experiencing gravity and experiencing uniform acceleration. The effects are locally identical. This is not merely an analogy; it's a fundamental assertion about gravity's nature and its intimate connection to spacetime's geometry.

Einstein recognized that if gravity and acceleration truly have equivalent effects, then gravity must possess properties similar to acceleration. Since acceleration is effectively nullified in a free-falling frame of reference (like astronauts in orbit feeling weightless because they are constantly falling around Earth), this suggested that gravity could also be nullified by finding the correct reference frame. This led Einstein to propose that gravity is not a force in the traditional Newtonian sense but rather a manifestation of spacetime's curvature.

In a uniformly accelerating elevator, the sensation of being pressed against the floor is a direct result of the elevator's upward motion through space. Every object inside, including you and the ball, is being pushed upwards by the elevator's floor. An observer outside the elevator would see the ball continuing on a straight path while the elevator accelerates upwards to meet it. However, for the person inside, it appears as though an invisible force is pulling the ball downwards.

This equivalence between gravity and acceleration is the crucial link connecting the geometric picture of spacetime curvature to gravity's observable effects. If gravity can be mimicked by acceleration, and acceleration is a consequence of motion through spacetime, then gravity must also be related to spacetime's geometry. Einstein realized that the presence of mass and energy causes spacetime to curve, and this curvature dictates objects' motion. What we perceive as the "force" of gravity is, in reality, objects following the "straightest possible paths"—geodesics—through this curved spacetime.

Consider the accelerating elevator again. As it accelerates upwards, the floor constantly moves to catch up with objects inside. If a light beam were shone horizontally across the elevator, an observer inside would see it appear to curve downwards. This is because, while the light travels in a straight line through space at a constant speed, the elevator itself accelerates upwards during the light's transit time. From the elevator's perspective, the destination on the opposite wall has moved upwards, making the light's path appear bent.

The Equivalence Principle suggests that a similar effect would occur in a gravitational field. If you were in a gravitational field and shone a light beam horizontally, you would expect it to bend downwards, just as in the accelerating elevator. This was a profound prediction: that gravity itself could bend light. This phenomenon, known as gravitational lensing, was one of General Relativity's first and most dramatic confirmations. During a 1919 solar eclipse, observations of starlight passing near the Sun revealed a slight but measurable deflection, precisely as Einstein had predicted. The Sun's mass had curved the spacetime around it, causing light from distant stars, traveling along geodesics in this curved spacetime, to bend from its original path.

The Equivalence Principle illuminates why objects of varying masses fall at the same rate in a gravitational field, a phenomenon that puzzled ancient Greek scientists and was famously demonstrated by Galileo's Leaning Tower of Pisa experiments (even if apocryphal). When a heavy cannonball and a light musket ball are dropped from the same height, they hit the ground simultaneously, assuming no air resistance. Newtonian physics explains this universal free fall by noting that while more massive objects experience a stronger gravitational force, they also possess greater inertia, requiring more force to accelerate. These two effects precisely counteract each other, resulting in identical acceleration for all objects.

General Relativity offers a more fundamental explanation. The Equivalence Principle posits that experiencing a gravitational field is indistinguishable from being in a uniformly accelerating frame of reference. This implies that the motion of objects in a gravitational field is not due to an active force, but rather to the geometry of spacetime itself. All objects, irrespective of mass or composition, follow the same paths (geodesics) in a given gravitational field. The perceived "force" of gravity is simply a consequence of these objects traversing the curved paths dictated by spacetime geometry.

Consider the accelerating elevator again: a dropped ball falls, while a helium balloon rises. This isn't due to gravity acting differently on them, but rather the air within the elevator. The accelerating elevator pushes upwards on the air, and the air, in turn, pushes upwards on both the ball and the balloon. The air pressure gradient causes apparent buoyancy. The denser ball is "pushed" downwards by the air pressure, while the less dense balloon is pushed upwards.

The situation in a gravitational field is analogous. Earth's gravity warps spacetime, and air, like all matter, follows geodesics in this warped spacetime. As a fluid, air distributes itself within the gravitational field. Denser objects, like the ball, move towards regions of greater spacetime curvature, while less dense objects, like the balloon, appear to move "upwards" relative to the denser surrounding medium, which we perceive as buoyancy. The Equivalence Principle helps us understand that these effects, seemingly due to different gravitational interactions, are actually consequences of how objects move through universally curved spacetime.

The implications of the Equivalence Principle extend to the very nature of gravitational fields. If gravity is indistinguishable from acceleration, then a gravitational field is not a static entity but a feature of spacetime's dynamics. In a freely falling frame, gravity effectively vanishes, meaning it's a property of motion within spacetime, not of space itself. This insight is critical for understanding phenomena like black holes.

Near a black hole, spacetime is so intensely curved that geodesics are severely distorted. The Equivalence Principle clarifies why escape is impossible once an object crosses the event horizon. Inside the event horizon, all spacetime paths, all geodesics, point inevitably towards the singularity at the center. It's not about needing an impossibly powerful rocket; it's that the very geometry of spacetime dictates that all possible paths lead inward. Imagine being in a river flowing increasingly fast towards a waterfall; if the current exceeds your swimming speed, you'll be carried over. With a black hole, the "current" of spacetime itself is so strong that it overwhelms even the speed of light, drawing everything inward.

Mercury's subtle orbital precession, a long-standing puzzle for Newtonian gravity, further testifies to the Equivalence Principle's power. Mercury's orbit around the Sun isn't a perfect ellipse; its closest approach (perihelion) slowly shifts with each orbit. While highly successful, Newton's laws couldn't fully explain this anomaly. However, General Relativity, by incorporating spacetime curvature caused by the Sun's mass and describing Mercury's motion as a geodesic in this curved spacetime, precisely predicted the observed precession. The orbit is not an ellipse in flat space; it's a non-closed geodesic in curved spacetime.

The Equivalence Principle is not merely theoretical; it has practical consequences, forming the foundation of GPS satellite operation. Satellites in orbit move at high speeds and experience weaker gravity than on Earth. Both factors, predicted by General Relativity, affect the passage of time. GPS satellite clocks run slightly faster due to weaker gravity and slightly slower due to their speed. Accounting for these relativistic effects, rooted in the Equivalence Principle, is crucial for GPS accuracy; without them, positions would drift by several kilometers daily.

Essentially, the Equivalence Principle frees us from conceiving gravity as an active force. Instead, it reveals gravity as a passive consequence of the universe's shape. When we feel gravity's pull, we are simply experiencing our motion along the natural, straightest paths available in a spacetime molded by mass and energy. The accelerating elevator profoundly reminds us that the sensation of weight, the apparent force keeping us grounded, is intimately connected with the geometry of our cosmic motion. This principle, in its elegant simplicity, establishes a framework for understanding the universe not as a stage for forces, but as a dynamic, geometric entity dictating the motion of everything within it.

The elegance of General Relativity, born from abstract thought and profound mathematics, would remain a fascinating theoretical construct if not for its remarkable predictive power, consistently substantiated by observational evidence. Einstein's theory not only reframed our understanding of gravity but also offered testable predictions deviating from the century-old Newtonian paradigm, providing crucial experiments for validation. Subsequent decades have seen increasingly precise observations repeatedly confirming the accuracy of Einstein's geometric interpretation of gravity, solidifying its place as a cornerstone of modern cosmology.

General Relativity's most significant historical confirmation was its explanation of Mercury's anomalous perihelion precession. For decades, astronomers meticulously tracked planetary orbits, finding Newton's laws remarkably successful, yet a subtle discrepancy in Mercury's orbit persisted. The perihelion, the point closest to the Sun in Mercury's elliptical path, didn't return to the exact same position with each orbit. Instead, it shifted, or "precessed," by about 43 arcseconds per century, an amount not explained by other planets' gravitational influences. This minute but persistent discrepancy troubled Newtonian physics, and various hypotheses, including a hypothetical planet "Vulcan," proved unsatisfactory.

Einstein's General Relativity offered an immediate and elegant solution. In GR, gravity isn't a force across empty space but a manifestation of spacetime's curvature caused by mass and energy. The Sun, a massive object, warps spacetime around it. Mercury, in its orbit, doesn't follow a strictly elliptical path in a flat plane but rather traces a geodesic—the straightest possible path—through this curved spacetime. The cumulative effect of this spacetime curvature, particularly near the Sun, causes the slight, persistent precession of Mercury's perihelion. GR's mathematical predictions for this precession matched the observed 43 arcseconds per century with astonishing accuracy, a profound triumph that resolved long-standing astronomical puzzles that Newtonian mechanics couldn't explain.

Another monumental validation of General Relativity came from observing the bending of starlight by the Sun's gravity. Einstein's theory predicted that both massive objects and light itself, despite having no rest mass, would be affected by spacetime curvature. This effect, termed gravitational lensing, suggested that as light from a distant star passed near a massive object like the Sun, its path would deflect, causing the star's apparent position to shift. Einstein calculated this deflection to be precisely twice the amount predicted by Newtonian physics, which considered light as particles (photons) pulled by gravity.

The opportunity to test this prediction arose during a total solar eclipse on May 29, 1919. Under Sir Arthur Eddington's guidance, expeditions to Sobral, Brazil, and Príncipe, off West Africa, observed stars visible around the Sun during totality. As the Moon's shadow swept across Earth, starlight passing near the Sun's limb was captured on photographic plates. By comparing these stars' positions with their known positions when the Sun was absent, astronomers measured any apparent shift. The results, announced in November 1919 at a joint meeting of the Royal Society and the Royal Astronomical Society in London, were electrifying. The observed starlight deflection closely matched GR's predicted value, confirming the bending of light by gravity and dramatically demonstrating the theory's validity. This confirmation was widely reported, catapulting Einstein to international fame and solidifying General Relativity's position as the leading theory of gravity.

Since then, gravitational lensing has been observed with even greater precision in numerous other contexts. Distant galaxies and quasars, whose light passes through the gravitational fields of intervening galaxies or galaxy clusters, exhibit distorted images, multiple images, or arcs—all direct consequences of spacetime curvature. These gravitational lenses act like cosmic telescopes, magnifying and distorting light from more distant objects, enabling astronomers to study the universe in previously impossible ways. The patterns and degrees of distortion provide detailed information about the lensing objects' mass distribution, independently verifying General Relativity's principles.

While the bending of starlight and Mercury's orbital precession confirmed static or slowly changing gravitational fields, General Relativity also made a more dynamic prediction: the existence of gravitational waves. These are ripples in the fabric of spacetime itself, generated by the acceleration of massive objects, much like a disturbance on a pond's surface creates ripples in the water. Einstein predicted that the universe's most violent and energetic events, such as colliding black holes or neutron stars, would produce gravitational waves that propagate outwards at the speed of light.

For nearly a century, gravitational waves remained a theoretical concept due to their incredibly weak and challenging-to-detect signals. However, advancements in sensitive instrumentation spurred the scientific community to pursue direct observation. This pursuit led to the construction of sophisticated laser interferometers, such as the Laser Interferometer Gravitational-Wave Observatory (LIGO) in the United States and Virgo in Italy. These detectors operate by measuring minute changes in the lengths of their arms, caused by the passage of a gravitational wave.

The groundbreaking discovery occurred on September 14, 2015, when LIGO detected GW150914—a faint yet distinct pattern of spacetime distortion. This signal was identified as originating from the merger of two stellar-mass black holes, over a billion light-years away, and perfectly aligned with General Relativity's predictions for such an event. This marked the direct confirmation of gravitational waves and the dynamic nature of spacetime, ushering in a new era of astronomy. Since then, LIGO and Virgo have detected numerous gravitational wave events, including neutron star mergers that also produce observable electromagnetic radiation, offering a "multi-messenger" perspective on cosmic cataclysms. Detailed analysis of these signals provides unprecedented accuracy in probing black hole and neutron star properties, further testing General Relativity in extreme gravitational environments.

For instance, the precise waveforms of merging black holes allow for stringent tests of GR's predictions regarding their nature, such as the no-hair theorem and the properties of the event horizon.

The consistent observational evidence—including the subtle shift in Mercury's orbit, the bending of starlight, and the direct detection of gravitational waves—has solidified General Relativity's credibility. These phenomena are not mere curiosities but fundamental predictions stemming directly from the theory's core principles: the equivalence principle and the geometric interpretation of gravity as spacetime curvature. The universe's precise adherence to these predictions significantly validates the theory.

This empirical success extends beyond theoretical vindication, forming the foundation for our understanding of extreme astrophysical phenomena. With General Relativity robustly supported by observations in relatively "mild" gravitational environments, scientists can confidently extrapolate its principles to regions of exceptionally strong gravity, such as those around black holes and neutron stars. The ability to explain and predict the behavior of these enigmatic objects, from spacetime warping to gravitational wave emission during mergers, directly results from the theory's observational validation. Without the empirical grounding from Mercury's orbit, solar eclipses, and gravitational wave detections, such extrapolations would be far more speculative.

General Relativity also explains other subtle observed effects. For example, the 1959 Pound-Rebka experiment provided direct terrestrial confirmation of gravitational redshift, a phenomenon predicted by the Equivalence Principle. This experiment measured the frequency change of photons as they traveled upwards in Earth's gravitational field. Photons climbing out of a gravitational potential well lose energy and their frequency decreases (redshift), while those falling in gain energy and their frequency increases (blueshift). Using gamma rays and the Mössbauer effect, the Pound-Rebka experiment accurately measured this tiny frequency shift over just 22.5 meters, confirming that gravity affects the passage of time itself—a concept known as gravitational time dilation, a direct consequence of spacetime curvature.

These observational successes have far-reaching implications, confirming General Relativity's validity and highlighting its indispensable role in modern astrophysics and cosmology. Our understanding of the universe's evolution, large-scale structure formation, stellar behavior, and the very nature of space and time are deeply intertwined with its principles. Accurate prediction and observation of phenomena like gravitational lensing are crucial for measuring cosmological distances and understanding the distribution of dark matter, which constitutes a significant portion of the universe's mass.

Furthermore, General Relativity's success in explaining phenomena from planetary wobbles to black hole collisions has made it an indispensable tool for understanding the universe. Continued refinement of observational techniques and new instruments promise to further test Einstein's theory and potentially uncover new physics. Yet, after over a century of scrutiny, General Relativity remains a remarkably robust and accurate description of gravity, a testament to its brilliant conception and the power of empirical verification. The evidence from diverse astrophysical phenomena consistently paints a picture of spacetime as a dynamic, malleable entity, its geometry sculpted by matter and energy, governing the dance of celestial bodies and the propagation of light and gravitational waves.

CHAPTER 3

The Anatomy of a Black Hole

The evolution from a brilliant, life-giving star to the mysterious void of a black hole is a cosmic spectacle of immense proportions. It's a testament to gravity's ultimate dominance. This transformation primarily occurs during the death of a massive star, a process spanning eons of nuclear fusion that culminates in a catastrophic collapse.

From their birth, stars are locked in a constant struggle: the outward pressure generated by nuclear fusion in their cores battles the inward pull of their immense gravity. For most of a star's existence, these forces achieve a delicate balance known as hydrostatic equilibrium, allowing stars like our Sun to shine steadily for billions of years.

However, this equilibrium is temporary. As a star depletes the nuclear fuel in its core (first hydrogen, then helium), it begins to fuse heavier elements. This process continues, producing progressively heavier elements such as carbon, oxygen, neon, silicon, and ultimately, iron. Iron is a critical element in stellar evolution because its fusion consumes energy rather than releasing it. When a star's core becomes predominantly iron, energy production ceases, shattering the core's equilibrium as the outward pressure from fusion disappears.

What follows is a rapid and dramatic implosion. The iron core, typically only a few thousand kilometers in radius, collapses under its own gravity in a fraction of a second. The extreme pressure forces electrons and protons to combine, forming neutrons and releasing a torrent of neutrinos. While these neutrinos carry away a significant amount of energy, the bulk of the gravitational potential energy is converted into a powerful outward shockwave. This shockwave blasts the star's outer layers into space in a spectacular explosion known as a supernova, one of the most energetic events in the universe, briefly outshining entire galaxies.

The fate of the stellar remnant after the supernova depends crucially on the mass of the collapsed core. If the core's mass is below approximately 1.4 times the mass of our Sun (the Chandrasekhar limit), the degeneracy pressure from the tightly packed neutrons will be enough to halt the collapse. This results in the formation of a neutron star—an incredibly dense object where matter is so compressed that a teaspoonful would weigh billions of tons. Neutron stars are fascinating in their own right, often spinning rapidly and emitting beams of radiation, appearing to us as pulsars.

However, if the remnant core's mass exceeds a critical limit, roughly two to three times the mass of the Sun (known as the Tolman-Oppenheimer-Volkoff limit, a value still being precisely determined due to uncertainties in the equation of state for neutron star matter), even neutron degeneracy pressure cannot withstand gravity's overwhelming force. In such cases, the collapse continues unchecked. There is no known natural force capable of stopping this relentless gravitational descent. The core collapses inward, and as it shrinks, its gravitational field intensifies, severely warping the fabric of spacetime.

As the core collapses, its density and the curvature of spacetime around it increase dramatically. The concept of escape velocity becomes paramount. For any object, there's a speed required to overcome its gravitational pull. For Earth, this is about 11 kilometers per second. For a collapsing stellar core, as its radius shrinks while its mass remains constant, the escape velocity at its surface increases. When this escape velocity reaches the speed of light, even light itself cannot escape, and at this point, a black hole is formed.

The event horizon marks the theoretical boundary from which nothing, not even light or gravitational waves, can escape a black hole. Anything that crosses this point is inevitably drawn toward the singularity at the black hole's center. This singularity is a point of infinite density and spacetime curvature where the known laws of physics, as described by General Relativity, break down. A complete understanding of the singularity likely requires a theory of quantum gravity, a field still under development.

Black holes form primarily from the collapse of massive stars, a direct consequence of Einstein's General Relativity. This theory states that mass and energy warp spacetime, which in turn dictates the movement of objects. In a stellar collapse, the concentration of mass becomes so extreme that spacetime is severely distorted. The event horizon represents the point where this distortion is so profound that all paths lead inward to the singularity.

While stellar collapse is common, it's not the only way black holes form. Most large galaxies, including the Milky Way, host supermassive black holes with masses millions or billions of times that of our Sun. Their formation is an active research area, with leading theories suggesting origins from the collapse of enormous gas clouds in the early universe or the merger of smaller black holes over cosmic time. The dense stellar environments at galactic centers could also contribute to their growth. These supermassive black holes significantly influence the dynamics, star formation, and overall evolution of their host galaxies. Understanding their formation is crucial for a comprehensive view of cosmic evolution, linking stellar birth to galactic architecture.

The transformation of a star into a black hole highlights the immense power of gravity and the finite lifespan of stars. It represents a cosmic recycling process where stellar remnants become the seeds for extreme gravitational phenomena, profoundly shaping the universe. The existence of black holes, predicted by theory and confirmed by observation, provides opportunities to explore the most extreme cosmic environments and expand our understanding of physics.

At the heart of a black hole, where our fundamental understanding of physics falters, lies the singularity. This isn't a conventional place but a theoretical point of infinite density and curvature at the black hole's core—the ultimate destination for anything crossing the event horizon. Imagine compressing the mass of an entire collapsed star, or even a colossal collection of stars, into an infinitesimally small volume. This illustrates the singularity, a point where matter and energy are compressed beyond comprehension, infinitely warping the fabric of spacetime.

In Einstein's General Relativity, a singularity is the inevitable outcome of extreme gravitational collapse. When a star or other object collapses under its own gravity, its radius shrinks, and if it surpasses the Tolman-Oppenheimer-Volkoff limit, no force can stop this process. This leads to an unbound increase in the density of matter and an increasingly extreme curvature of spacetime. The singularity represents the ultimate endpoint: a point where density and gravitational fields become infinite, and the equations of General Relativity break down, effectively rendering space and time incomprehensible.

To visualize this, imagine a heavy ball creating a deep dip in a stretched rubber sheet, representing spacetime curvature. If this ball were to shrink, becoming infinitely dense and pressing down with infinite force, the dip would become infinitely deep, tearing or stretching the sheet to an infinite extent. Similarly, within a black hole's event horizon, all paths lead inexorably towards this central point of infinite curvature—a one-way journey into an ultimate gravitational abyss.

The singularity isn't a physical surface but a conceptual boundary where General Relativity's mathematical description of gravity reaches its limits. All mass that falls into a black hole is theorized to concentrate at this dimensionless point. Objects approaching the singularity experience extreme tidal forces, a process known as "spaghettification," where the differential gravitational pull tears them into long, thin strands before they reach the singularity.

It's crucial to note that our current understanding of black holes through General Relativity is classical and does not incorporate quantum mechanics, which governs behavior at extremely small scales. At the singularity, quantum effects are expected to dominate, but we lack a unified theory of quantum gravity to describe what truly happens. Thus, the singularity remains a placeholder, marking the limits of our current knowledge.

Physicists speculate about various solutions to the singularity problem. Some theories suggest that quantum gravity might prevent infinite density, perhaps creating "fuzzballs" of exotic matter or a granular spacetime structure. Other ideas, like those from string theory, propose that the singularity might be replaced by a region of extremely high but finite density, or even a "bounce" into another spacetime or universe. However, these ideas are currently speculative, awaiting further theoretical development and observational evidence.

The concept of a singularity within General Relativity is intricate. For a non-rotating, uncharged black hole (Schwarzschild metric), the singularity is a point at a radial coordinate of r=0. However, in a rotating black hole (Kerr metric), the singularity is predicted to be a ring at the center, demonstrating that its precise nature may vary with the black hole's properties like spin. Despite these differences, the fundamental issue of infinite density and curvature remains a core prediction of the theory.

The event horizon conceals the singularity, leading to the "cosmic censorship" principle, proposed by Roger Penrose. This principle suggests the universe "protects" external observers from singularities by always shielding them with an event horizon. This containment prevents the breakdown of physical laws at the singularity from chaotically affecting the rest of the universe. If "naked singularities"—those not hidden by an event horizon—were common, the predictability of physics could be compromised. Cosmic censorship implies that the extreme conditions leading to a singularity are always contained, preserving the predictable causal structure of spacetime outside the event horizon.

The "infinite density" of a singularity signifies a point where our current theories, known to be incomplete at such scales, break down. Just as quantum mechanics limits the absolute precision of an electron's position, at the singularity, the extreme scales render the concepts of position and distance meaningless.

The singularity is a testament to General Relativity's power and elegance, yet it also marks the limit of its predictive capabilities. It represents a frontier in physics, signaling the need for a deeper, more comprehensive understanding of gravity and the fundamental nature of spacetime. While the mathematical conditions leading to a singularity can be described, what actually exists at that central point remains a profound mystery in astrophysics and theoretical physics. The study of black holes, therefore, goes beyond understanding these exotic objects; it involves pushing the boundaries of our knowledge and seeking the next revolution in our cosmic understanding. The event horizon, a veil hiding this ultimate mystery from direct observation, is a direct consequence of the extreme gravitational forces culminating in the singularity. Understanding the singularity is crucial not only for theoretical purposes but also for comprehending the ultimate fate of matter in the universe and the very nature of gravity.

The notion of "infinite density" challenges our intuition, which is based on a universe governed by well-understood physical laws. Within a singularity, these laws are expected to be drastically different. General Relativity describes gravity as spacetime curvature caused by mass and energy. As mass compresses into an ever-smaller volume, this curvature intensifies, becoming infinite at the singularity. This infinite warping of spacetime implies that the usual concepts of space, time, distance, and duration cease to be meaningful.

To grasp the immense forces involved, consider an astronaut falling into a black hole. As they near the singularity, the gravitational pull difference between their feet and head would become immense, stretching them like spaghetti—a phenomenon known as "spaghettification." For a stellar-mass black hole, these tidal forces would be lethal even before reaching the event horizon. However, for supermassive black holes, the event horizon is much larger, and the tidal forces at the horizon itself are much weaker. An astronaut falling into a supermassive black hole might cross the event horizon without immediately noticing anything catastrophic, continuing their journey toward the center, where they would experience increasingly extreme tidal forces until finally reaching the singularity.

The singularity, a theoretical concept within General Relativity, derives its characteristics from the theory's mathematical framework. For non-rotating, uncharged black holes (Schwarzschild black holes), the singularity is a central point where the radial coordinate becomes zero. However, in rotating black holes (Kerr black holes), the singularity is predicted to be a ring located in the equatorial plane. This difference arises because rotation introduces angular momentum, altering the spacetime structure around the black hole. Even with a ring singularity, both density and curvature are still projected to become infinite along the ring.

A key challenge for physicists is that General Relativity is a classical theory of gravity. It does not account for quantum mechanical effects, which are anticipated to be significant at the extremely small scales and high densities associated with singularities. Quantum mechanics describes the universe in terms of probabilities and discrete quantities, suggesting that a quantum theory of gravity might resolve the singularity by preventing infinite density. For instance, some theories propose a granular structure for spacetime at its smallest scales, which would prevent infinite compression. Other theories suggest that quantum effects could generate a repulsive force at extremely high densities, halting the collapse.

The inherent inaccessibility of singularities poses a major challenge to their study. The event horizon acts as a perfect one-way barrier, preventing any information, including light, from escaping. This means direct observation of the singularity or testing theories about its nature using electromagnetic radiation is impossible. Our understanding of the singularity is therefore largely based on theoretical extrapolations from General Relativity and indirect inferences drawn from observing the behavior of matter and light near black holes.

The concept of the singularity also raises profound questions about the fundamental nature of space and time. In classical physics, space and time are viewed as a continuous background for events. However, at the singularity, the extreme curvature of spacetime suggests that these concepts may break down. Some quantum gravity theories propose that spacetime itself is not fundamental but rather emerges from more basic quantum constituents. In such a framework, the singularity might represent a region where these underlying constituents are arranged in a fundamentally different way than the spacetime we experience.

The singularity serves as a reminder of the incompleteness of our current understanding of the universe. While General Relativity has been remarkably successful in describing gravity and black holes, a more complete theory, one that unifies gravity with quantum mechanics, is likely necessary to fully comprehend what occurs at the heart of a black hole. The pursuit of this quantum theory of gravity is one of the most active and exciting areas of research in theoretical physics. It is a quest to reconcile the large-scale, smooth fabric of spacetime described by Einstein with the strange, probabilistic world of quantum mechanics. Consequently, the singularity acts as a crucial focal point in this ongoing endeavor, pushing the boundaries of our scientific imagination and our ability to understand the most extreme phenomena in the cosmos. The very existence of a point of infinite density underscores the limitations of our current models and the need for a deeper, more fundamental understanding of the universe.

The event horizon, not the singularity, defines the true boundary of a black hole—the point of no return. It acts as a gateway, a cosmic turnstile, marking the precise limit beyond which escape is impossible. Imagine it as the edge of an infinitely deep well, where gravity's pull is so overwhelming that nothing, not even light, can break free.

This boundary is a direct consequence of Einstein's theory of General Relativity. When mass is sufficiently concentrated, spacetime warps severely, causing all paths, for both matter and light, to converge inward. The event horizon is the collection of points in spacetime from which a light signal, traveling at its fastest possible speed, can just barely reach infinity. Closer than this, even light is inexorably drawn toward the center. Essentially, it's the surface where the escape velocity equals the speed of light, ensuring that anything crossing this boundary is permanently trapped.

The size of this critical boundary is directly proportional to the black hole's mass. This relationship is quantified by the Schwarzschild radius, named after Karl Schwarzschild, who first derived this solution to Einstein's field equations. For a non-rotating, uncharged black hole (a Schwarzschild black hole), the event horizon is a perfect sphere, with its radius given by the formula $R_s = 2GM/c^2$, where G is the gravitational constant, M is the mass of the black hole, and

c is the speed of light. This formula reveals that a more massive black hole will have a larger event horizon.

For instance, a black hole with the mass of our Sun would have a Schwarzschild radius of approximately 3 kilometers (about 1.86 miles). If the Sun collapsed into a black hole, its event horizon would extend to this radius. In contrast, supermassive black holes found at the centers of galaxies, such as Sagittarius A* at the heart of the Milky Way, with masses millions of times that of the Sun, have correspondingly vast event horizons, spanning millions of kilometers. Sagittarius A*'s event horizon radius is roughly 12 million kilometers, a substantial but still minuscule size on galactic scales. This proportionality means that while stellar-mass black holes are relatively small and dense, supermassive black holes can have event horizons comparable in size to our solar system, though still much smaller than the stars within it.

The event horizon is often described as a "one-way membrane." This crucial concept means that for anything attempting to escape, crossing the event horizon is like stepping through a door with no handle on the other side, leading to a room with permanently sealed exits. Information, energy, and matter can fall into a black hole across this boundary, but nothing can ever emerge. This property is what makes black holes so enigmatic and powerful—they are ultimate traps, consuming everything that ventures too close without any possibility of release.

The nature of the event horizon is subtle. It's not a physical surface like a planet's surface that you could touch or stand on. Instead, it's a boundary in spacetime defined by the gravitational field. An observer falling into a black hole would not necessarily feel anything special as they crossed the event horizon. According to General Relativity, if the black hole is massive enough, the tidal forces at the horizon might be relatively weak. An astronaut falling into a supermassive black hole might, for a brief moment, feel no different from being in freefall in empty space, continuing their journey inward, unaware of the precise moment they passed the point of no return.

From the perspective of a distant observer, an astronaut falling into a black hole would appear to experience a dramatic slowing of time as they approach the event horizon. This extreme warping of spacetime would cause their image to dim and redden due to gravitational redshift, eventually seeming to freeze at the horizon, never quite crossing it. This phenomenon, known as time dilation, illustrates how time passes more slowly in stronger gravitational fields, making the astronaut's arrival at the horizon appear to take an infinite amount of time to the distant observer.

This discrepancy in temporal experience underscores the relativity of spacetime. While the astronaut experiences crossing the event horizon as a finite event, it is an asymptotic process for the distant observer. This difference is a direct consequence of how gravity, as described by General Relativity, impacts the passage of time. The event horizon is not a local property but a global feature of spacetime geometry, defined by the collective gravitational pull that makes escape impossible.

The event horizon is crucial to many theoretical concepts surrounding black holes, including the proposed information paradox. The idea that information about matter falling into a black hole is permanently lost beyond the event horizon contradicts a fundamental principle of quantum mechanics, which states that information cannot be destroyed. Stephen Hawking demonstrated that black holes emit Hawking radiation, a thermal spectrum of particles that causes them to slowly evaporate. However, the connection between this radiation and the information that fell into the black hole remains a complex and unresolved puzzle, with the event horizon acting as the barrier that obscures this information.

Furthermore, the event horizon marks a fundamental alteration in the causal structure of spacetime. Inside the event horizon, all future light cones (the paths light can take) point towards the singularity, meaning every possible future for an object within this boundary is directed towards the center, rendering outward movement impossible. Outside the event horizon, despite strong gravity, light and matter can still escape with sufficient velocity. The event horizon is the precise boundary separating these two fundamentally different regimes of spacetime.

The size of the event horizon, determined by the Schwarzschild radius, is inextricably linked to the black hole's mass. While mass changes very slowly over cosmic timescales due to accretion and Hawking radiation, these changes are minuscule. Thus, for practical observational purposes, a black hole's event horizon can be considered a fixed boundary. However, any change in the black hole's mass would proportionally alter the size of its event horizon, highlighting the dynamic relationship between mass, gravity, and the structure of spacetime.

The concept of the event horizon is so central to the definition of a black hole that the black hole itself is often described as the region of spacetime enclosed by this boundary, rather than just the singularity. Without this boundary, the object would merely be an extremely dense star or an exotic compact object, not a true black hole. The event horizon imparts the characteristic "blackness" to these celestial bodies, acting as a perfect, one-way cosmic trap from which no light can escape. Its size is dictated by the very mass that creates it, an elegantly simple yet profoundly consequential relationship that defines the cosmic scale of these enigmatic entities and shapes our understanding of the universe's most extreme gravitational phenomena.

The cosmos is home to a diverse array of celestial objects, among which black holes are particularly extreme and enigmatic. Though all black holes share the fundamental characteristic of gravity so potent that not even light can escape, they vary significantly. They are primarily categorized by mass and cosmic epoch of birth, broadly classified into three main types: stellar, supermassive, and the theoretical primordial black holes. Each type offers unique insights into the universe's processes, from stellar deaths to its nascent moments. Understanding these variations is crucial, especially when considering the captivating hypothesis that a black hole might have birthed our universe.

Stellar black holes are the most common type, formed from the dramatic demise of massive stars. When a star, at least 20 to 25 times the mass of our Sun, exhausts its nuclear fuel, it can no longer counteract the inward pull of gravity. This triggers a catastrophic gravitational collapse. The star's core implodes, compressing its matter into an incredibly small volume. This implosion can lead to a supernova, but if the remnant core's mass exceeds the Tolman-Oppenheimer-Volkoff (TOV) limit (roughly 2 to 3 solar masses), even neutron degeneracy pressure cannot halt the collapse. The core collapses past the point of forming a neutron star, and gravity reigns supreme, crushing all matter into a singularity and forming a stellar black hole.

Stellar black holes typically range from 5 to a few tens of solar masses. However, through accretion—gravitationally pulling in gas and dust from their surroundings—some can grow significantly larger. In binary star systems, a stellar black hole can siphon material from its companion star, forming a superheated accretion disk that emits intense X-rays. These X-ray binaries are a primary method astronomers use to detect and study stellar black holes. Stellar black holes can also merge with other stellar objects, including other black holes, leading to even more massive black holes. Gravitational wave signals detected by observatories like LIGO and Virgo provide direct evidence of these energetic mergers, confirming their existence and offering insights into their population dynamics.

Supermassive black holes (SMBHs) represent an entirely different order of magnitude. These behemoths reside at the centers of most, if not all, large galaxies, including our Milky Way, with masses ranging from hundreds of thousands to billions of times that of our Sun. The precise mechanisms of SMBH formation and growth remain an active area of research. Several theories exist, and it's likely a combination of processes contributes to their colossal growth.

One prominent theory suggests that SMBHs begin as "seed" black holes, possibly formed from the collapse of the very first, massive stars (Population III stars) in the early universe, or directly from the gravitational collapse of massive gas clouds. Once these seeds are in place, they grow over billions of years, primarily through the accretion of gas and stars from the surrounding galactic environment. As material falls towards the galactic center, it can form an accretion disk, similar to those around stellar black holes but on a much grander scale. This infalling matter, especially abundant in the early universe, likely fueled rapid growth.

Supermassive Black Hole (SMBH) Growth: Galaxy mergers are a key mechanism. When galaxies collide, their central SMBHs spiral together and merge, significantly increasing the mass of the resulting SMBH. This common cosmic event is a major contributor to SMBH growth over time. Evidence for this comes from observations of Active Galactic Nuclei (AGN) and quasars, which are powered by actively accreting SMBHs. The intense radiation and jets from these objects demonstrate the immense gravitational power of the central black hole.

SMBH Mass Range and Size: SMBHs exhibit a vast mass range, from approximately 100,000 to over 10 billion solar masses. Their event horizons are correspondingly enormous. For example, Sagittarius A* (the Milky Way's SMBH) has a mass of about 4 million solar masses and a Schwarzschild radius of roughly 12 million kilometers (17 times the Sun's radius). A 10 billion solar mass black hole would have an event horizon roughly the size of our solar system out to Neptune. These immense sizes allow SMBHs to significantly influence host galaxy dynamics, impacting star formation, stellar motion, and overall galaxy evolution. The SMBH at the center of M87, famously imaged by the Event Horizon Telescope, has a mass of about 6.5 billion solar masses and an event horizon spanning nearly 25 billion kilometers.

Primordial Black Holes: This speculative third category of black holes is theorized to have formed not from stellar collapse, but from the extreme densities and pressures in the very early universe, shortly after the Big Bang. During this incredibly hot and dense epoch, matter might have become so compressed in some regions that it collapsed directly into black holes.

Formation of Primordial Black Holes: Unlike white dwarfs and neutron stars, whose formation is governed by the Chandrasekhar or TOV limits, primordial black holes would not have been constrained by these limits. Their mass could have spanned an incredibly wide range, from asteroid-sized to thousands of solar masses, due to density fluctuations in the early universe, which could have been large enough in localized regions to directly form black holes.

Significance of Primordial Black Holes: The existence of primordial black holes is unconfirmed, and their properties remain theoretical, but they are a compelling subject of study for several reasons. Firstly, they offer a potential explanation for the missing mass in the universe, as they could constitute a significant portion of dark matter. As black holes do not interact electromagnetically, they would be invisible to most telescopes, and their gravitational influence would be the primary way they affect their surroundings. If primordial black holes make up a substantial fraction of dark matter, it would resolve a long-standing cosmological puzzle. Secondly, the theoretical framework for their formation aligns with our understanding of the Big Bang and early universe physics. Inflationary cosmology, a period of rapid expansion in the first moments after the Big Bang, could have generated the necessary density perturbations for primordial black hole formation. Their abundance and mass distribution would depend on the specifics of the inflationary model and the nature of the early universe.

The study of primordial black holes (PBHs) offers crucial insights into the hypothesis that a black hole may have been the progenitor of our universe. If the Big Bang originated within a black hole, then PBHs, formed in the extreme conditions of the early universe, represent the closest analogues to such an object that we can currently conceive. This highly speculative concept suggests that a pre-existing black hole, possibly a very large one, could have undergone a process leading to the rapid expansion of spacetime, effectively giving birth to our universe. While this idea involves complex theoretical physics, including quantum gravity and the nature of singularities, exploring PBH properties and formation mechanisms provides a conceptual bridge to understanding such cosmic origins.

Should PBHs exist, their characteristics could offer clues about the conditions of the incredibly young universe and potentially illuminate the ultimate beginnings of existence. The ongoing search for evidence of PBHs through observational channels like gravitational lensing, gamma-ray bursts, and their potential contribution to gravitational wave backgrounds, is an active and exciting frontier in modern astrophysics. Their presence or absence would have profound implications for our understanding of cosmology, dark matter, and the fundamental nature of gravity.

The distinction between stellar, supermassive, and primordial black holes underscores the universe's incredible dynamism and the diverse paths that matter and gravity can take. Stellar black holes are the dramatic end-products of stellar evolution—remnants of cosmic violence. Supermassive black holes act as the gravitational anchors of galaxies, shaping their evolution over billions of years. Primordial black holes, if they exist, are whispers from the universe's infancy, potentially linking us to the very moments of creation. Each type, with its unique mass, formation history, and implications, contributes to our ever-evolving picture of the cosmos and its profound mysteries, including the audacious idea that our universe might have emerged from the heart of a black hole.

The event horizon, the boundary of a black hole, marks an ultimate point of no return. It's not a physical surface but a one-way membrane in spacetime. Once an object or photon crosses this invisible threshold, it is inexorably drawn towards the singularity at the black hole's center. The experience of crossing this cosmic precipice, as theoretical physics suggests, would be profoundly alien and utterly destructive for any macroscopic object.

Consider an astronaut bravely venturing towards a stellar-mass black hole. As the spacecraft approaches the event horizon, the gravitational pull becomes overwhelmingly strong. The immediate sensation, however, might be stretching rather than crushing. This peculiar elongation is a direct consequence of tidal forces, which arise because the black hole's gravitational pull isn't uniform across the spacecraft. The part of the spacecraft closer to the black hole experiences a significantly stronger pull than the part farther away. This differential force acts like a cosmic vise, stretching the object along the direction of the black hole while simultaneously compressing it from the sides.

This phenomenon is graphically known as "spaghettification." For our hypothetical astronaut, the gravitational gradient across their body would become so extreme that their feet, if pointed towards the black hole, would be pulled with far greater force than their head, leading to rapid and irreversible elongation. Simultaneously, immense gravity would squeeze them inwards, flattening them like a noodle. The greater the black hole's mass, the more gradual this stretching effect would be at the event horizon. For supermassive black holes, tidal forces at the event horizon are relatively weak, potentially allowing an astronaut to cross without immediately experiencing violent spaghettification. However, for a stellar-mass black hole, tidal forces become insurmountable much further out, ensuring spaghettification begins long before the horizon itself.

The experience of falling into a black hole differs dramatically based on the observer's perspective. From the falling astronaut's viewpoint, they would perceive themselves crossing the event horizon. Time, for them, would flow normally, and they might even observe the universe outside continuing its cosmic dance, albeit heavily distorted by the black hole's gravity. Light from distant stars would appear bent around the black hole, creating warped and duplicated images. Their journey towards the singularity would be irreversible; from their perspective, they would simply continue to fall inwards, their spaghettified form eventually succumbing to the singularity's immense gravity.

Gravitational time dilation creates a stark contrast in the experience of an astronaut falling into a black hole, as viewed by a distant, orbiting observer. From this external perspective, the astronaut's clock would appear to slow down as they near the event horizon. The light they emit would exhibit increasing redshift, its wavelength stretched by the intense gravity. Consequently, the astronaut would seem to decelerate and effectively freeze in time, gradually fading into invisibility as their light becomes infinitely redshifted. They would never actually appear to cross the event horizon, but rather dim and vanish over an infinite duration, a direct consequence of Einstein's theory of general relativity, which dictates that time slows in stronger gravitational fields. Thus, from an external viewpoint, the event horizon acts as an uncrossable boundary within a finite amount of external observer's time.

The singularity, a theoretical point of infinite density at a black hole's core, remains a profound mystery in physics. General relativity predicts that all matter falling into a black hole is crushed into this single point. However, general relativity is known to fail under such extreme conditions, where quantum effects are expected to become dominant. A complete understanding of the singularity requires a theory of quantum gravity, a unified framework that reconciles general relativity with quantum mechanics—a theory yet to be fully developed.

It's crucial to acknowledge that these descriptions stem from our current theoretical understanding and extrapolations based on Einstein's theory of general relativity. Direct observation of an object crossing a black hole's event horizon is impossible, as nothing can escape from within to relay its experiences. Nevertheless, the theoretical framework offers a compelling, albeit terrifying, glimpse into the ultimate fate of matter succumbing to a black hole's irresistible pull. The extreme tidal forces, the warping of spacetime, and the divergent perspectives of falling and distant observers depict an environment so alien that it challenges our everyday intuition about space, time, and gravity. The very notion of "falling in" is not a simple descent, but a complex interplay of forces and relativistic effects culminating in the complete dissolution of form and identity within the enigmatic depths of the black hole.

CHAPTER 4

The Big Bang: Birth of Our Universe

The universe we observe today, with its vast array of galaxies, stars, and planets undergoing constant evolution, did not appear in its current form instantaneously. Instead, scientific understanding, supported by extensive observational data and theoretical frameworks, points to a singular, cataclysmic origin: the Big Bang theory, a cornerstone of modern cosmology. This theory proposes that approximately 13.8 billion years ago, the observable universe, along with space and time, emerged from an infinitesimally small, unimaginably hot, and incredibly dense state, often referred to as a singularity. This was not an explosion *in* space, but rather an expansion *of* space, much like points on the surface of an inflating balloon moving farther apart as the balloon itself stretches.

Following this initial rapid expansion, the universe began to cool as its energy density decreased and its temperature dropped. In its earliest moments, the universe was a searing "primordial soup" of fundamental particles and energy. Protons and neutrons, the building blocks of atomic nuclei, formed from quarks and gluons. As the temperature continued to fall, a crucial event occurred: the formation of light atomic nuclei, primarily hydrogen and helium, through a process called Big Bang nucleosynthesis. This period, lasting only a few minutes, is remarkably well-understood. The predicted relative abundances of hydrogen, helium, and trace amounts of lithium align precisely with observations, providing compelling evidence for the Big Bang model.

For hundreds of thousands of years after the Big Bang, the universe remained opaque. The intense heat prevented electrons from binding to atomic nuclei, creating a plasma. Photons (light particles) constantly scattered off these free electrons, unable to travel freely, effectively making the universe invisible. However, as the universe continued to expand and cool, a pivotal transition occurred. When the temperature reached approximately 3,000 Kelvin, electrons combined with atomic nuclei to form stable, neutral atoms—a process known as recombination. This dramatically reduced photon scattering, allowing light to travel unimpeded across the universe, marking the moment the cosmos became transparent.

The photons released during this era of recombination have been traveling through space ever since, carrying information about the early universe's conditions. Over billions of years of cosmic expansion, these photons have stretched to longer wavelengths, shifting from visible light into the microwave portion of the electromagnetic spectrum. This relic radiation, known as the Cosmic Microwave Background (CMB), is a nearly uniform glow of microwave radiation coming from all directions in the sky. It represents an all-pervading echo of the Big Bang, a snapshot of the universe when it was only a few hundred thousand years old. While the CMB exhibits remarkable uniformity, closer inspection reveals tiny temperature fluctuations, on the order of parts per hundred thousand. These minuscule variations, precisely observed by missions like COBE, WMAP, and Planck satellites, are of paramount importance to understanding the universe's evolution.

The universe we observe today, with its grand structures of galaxies, galaxy clusters, and superclusters, originated from subtle variations in the early universe known as anisotropies—deviations from perfect uniformity. These slightly denser regions, over immense timescales, were amplified by gravity. As more matter was drawn to these gravitational wells, gas and dust accumulated, eventually collapsing to form the first stars and galaxies. The precise characteristics of these cosmic microwave background (CMB) temperature fluctuations, meticulously mapped by advanced telescopes, provide crucial insights into fundamental cosmic parameters such as the universe's age, its composition (ordinary matter, dark matter, and dark energy), and its expansion rate.

The prevailing scientific explanation for the universe is the Standard Cosmological Model, often called the Lambda-

CDM (ΛCDM) model. This model builds upon the Big Bang theory and incorporates the discoveries of dark energy (Λ) and cold dark matter (CDM). It effectively accounts for a wide range of cosmological observations, including the universe's expansion, the abundance of light elements, the CMB's properties, and the formation of large-scale structures. The ΛCDM model posits that the universe is composed of approximately 5% ordinary matter (visible stars, planets, and ourselves), about 27% dark matter, and a dominant 68% dark energy. Dark matter, invisible due to its lack of interaction with light, is detected through its gravitational influence on visible matter and radiation. Dark energy, a more enigmatic component, is believed to be responsible for the universe's accelerating expansion, a discovery that earned the Nobel Prize in Physics in 2011.

The universe's evolution from the Big Bang to its current state is a continuous narrative driven by fundamental physical laws. After the formation of the first atoms, the universe was primarily a quiescent expanse of neutral hydrogen and helium gas. Gravity gradually drew this gas into denser clouds, where increasing pressure and temperature at their cores eventually led to nuclear fusion, the process by which atomic nuclei combine to release vast amounts of energy. This heralded the birth of the first stars, which were likely more massive and short-lived than contemporary stars. These early stars forged heavier elements within their cores, dispersing them into the interstellar medium through supernova explosions.

These newly synthesized heavier elements became the essential building blocks for subsequent generations of stars and planets. As more stars emerged, gravity caused them to coalesce, forming the first galaxies. These early galaxies were smaller and less regular than the grand spiral and elliptical galaxies we observe today. Over cosmic time, they merged and interacted, growing in size and complexity, creating the dynamic "cosmic web" of galaxies, clusters, and filaments separated by vast, largely empty voids. The universe continues to expand, and the accelerating nature of this expansion, driven by dark energy, suggests that galaxies will increasingly recede from one another. In the distant future, this could lead to a scenario where galaxies beyond our local group become causally disconnected, disappearing beyond our observable horizon.

Despite its remarkable success, the Standard Cosmological Model faces ongoing challenges and areas of active research. The precise nature of dark matter and dark energy remains unknown, presenting significant puzzles in fundamental physics. While their gravitational effects are observable, their underlying constituents and properties are still elusive. Furthermore, the Big Bang singularity itself, an initial point of infinite density, poses a theoretical hurdle. Current laws of physics, particularly general relativity, are inadequate under such extreme conditions. A complete understanding of the universe's very beginning will likely require a theory of quantum gravity, a theoretical framework that unifies quantum mechanics and general relativity, and is still under development.

The Big Bang theory, specifically the CDM model, offers a consistent and robust framework for understanding the universe's history and evolution, despite some lingering questions. A wealth of observational data—including the expansion rate, light element abundance, CMB anisotropies, and large-scale galaxy distribution—converges to paint a compelling picture of a universe that originated from a hot, dense state and has been expanding and evolving for billions of years. This detailed cosmic narrative, pieced together from ancient light and subtle gravitational forces, is a testament to human ingenuity and scientific inquiry. The Big Bang theory is a well-validated scientific model, forming the foundation of our cosmological understanding and the benchmark for evaluating new ideas.

The concept of a dynamic, expanding universe challenges the long-held belief in a static cosmos. For millennia, humanity viewed the celestial sphere as fixed and unchanging. However, advancements in observational techniques and scientific curiosity have revealed a universe in constant motion, a "grand cosmic ballet" orchestrated by the stretching of spacetime. This profound realization—that galaxies are systematically moving away from us—is supported by robust evidence, most notably the phenomenon of redshift and Hubble's Law. These discoveries fundamentally reshaped our understanding of the cosmos and provided crucial support for the Big Bang model.

To grasp this evidence, one must understand how light conveys information about the motion of its source. Light behaves as a wave, characterized by wavelength and frequency. Similar to ripples on a pond or sound waves from a siren, these properties change with the source's motion. A light source moving towards an observer compresses its waves, leading to shorter, bluer wavelengths (blueshift). Conversely, a source moving away stretches its waves, resulting in longer, redder wavelengths (redshift). The magnitude of this shift is directly proportional to the relative speed of the source. This Doppler effect, named after Christian Doppler, is a fundamental principle allowing astronomers to determine the radial velocity of celestial objects by analyzing their emitted light.

In the early 20th century, astronomers began meticulously analyzing light from distant galaxies. Between 1912 and 1914, Vesto Slipher observed that spectral lines from "spiral nebulae" (then thought to be gas clouds within our Milky Way) were systematically shifted towards the red end of the spectrum. This indicated that most of these nebulae were receding from Earth at remarkably high speeds, some reaching hundreds of kilometers per second. However, the exact nature and distances of these nebulae were unclear at the time, and the full implications of Slipher's findings—evidence for cosmic expansion—were not immediately understood.

The true significance of this widespread redshift was elucidated by American astronomer Edwin Hubble. Working at Mount Wilson Observatory in the 1920s with his assistant Milton Humason, Hubble embarked on measuring the distances to these spiral nebulae, a formidable task in astronomy. He employed various methods, notably identifying Cepheid variable stars in closer nebulae. Cepheids are pulsating stars whose pulsation period directly correlates with their intrinsic luminosity. By measuring a Cepheid's period, astronomers can determine its intrinsic brightness. Comparing this to its apparent brightness (as seen from Earth), they can then calculate its distance using the inverse square law.

Hubble's meticulous measurements of Cepheid variables in galaxies like Andromeda (initially called the Andromeda Nebula) were a turning point. His calculations showed that these nebulae were not within our Milky Way but were, in fact, entire galaxies millions of light-years away. This discovery alone was monumental, expanding our perception of the universe's scale and confirming our galaxy was but one of countless others. Hubble then combined his distance measurements with Slipher's redshift data and his own observations, plotting the recession velocity of each galaxy against its estimated distance.

This plotting revealed a clear and striking linear relationship: the farther away a galaxy was, the faster it appeared to be moving away from us. This relationship, famously known as Hubble's Law, is expressed mathematically as $v = H * d$, where 'v' is the recessional velocity, 'd' is the distance, and 'H' is the Hubble constant, representing the universe's expansion rate. The Hubble constant's value is crucial for determining the universe's age and scale. Early estimates by Hubble and Humason were around 500 kilometers per second per megaparsec (a megaparsec being approximately 3.26 million light-years), meaning a galaxy recedes about 500 km/s faster for every megaparsec farther away it is.

The implications of Hubble's Law were profound. It provided undeniable observational evidence of an expanding universe. The fact that almost all galaxies are receding from us, with their recession velocity proportional to their distance, is precisely what one would expect in a uniformly expanding universe. This can be visualized as points on an inflating balloon; every point moves away from every other point, with more distant points appearing to recede faster due to more expanding surface between them. This cosmic expansion is not an explosion into pre-existing space, but rather the stretching of space itself.

This observed expansion is a cornerstone of the Big Bang theory. If the universe is currently expanding, it logically follows that it was smaller, denser, and hotter in the past. Tracing this expansion backward in time suggests that all matter and energy in the observable universe originated from a single point of immense density and temperature—the primordial singularity. The redshift in light from distant galaxies is therefore direct evidence of this ongoing expansion, a cosmic echo of that initial, explosive genesis.

The discovery and verification of Hubble's Law was not a single event but the culmination of decades of work by numerous astronomers, built upon theoretical frameworks and technological advancements. The physics behind redshift measurement is rooted in spectroscopy, the study of how matter interacts with electromagnetic radiation. By dispersing light from celestial objects into its constituent wavelengths, astronomers create a spectrum—a unique "barcode." This spectrum contains absorption or emission lines at specific wavelengths, corresponding to particular elements. When these spectral lines are shifted from their known laboratory wavelengths, it indicates motion, with a greater shift signifying a greater speed.

The measurement of spectral shifts has undergone significant improvements over time. Early observations by Slipher were limited by technology, but Hubble and Humason achieved greater precision using advanced telescopes and spectrographic techniques. This precision was vital for establishing Hubble's Law and was supported by theoretical developments, particularly Einstein's 1916 theory of General Relativity, which reframed gravity as spacetime curvature caused by mass and energy.

In the early 1920s, Alexander Friedmann, an independent Russian mathematician and cosmologist, used Einstein's equations to derive solutions describing an expanding or contracting universe. Einstein initially resisted this, introducing a "cosmological constant" for a static universe, a move he later called his "biggest blunder" after Hubble's evidence emerged. Friedmann's work, along with that of Belgian priest and physicist Georges Lemaître, provided the theoretical basis for an expanding universe. Lemaître was the first to propose the Big Bang theory, suggesting the universe began as a primeval atom.

In 1927, Lemaître published his findings based on Friedmann's equations and observational data, two years before Hubble's seminal paper. He presented evidence for an expanding universe and estimated an expansion rate remarkably close to Hubble's later findings. However, his work, published in a less widely circulated French journal, initially had limited impact. Nevertheless, his theoretical insights were crucial for interpreting observed redshifts and the expanding universe.

Hubble's Law represents more than just measuring distances and redshifts; it's a testament to the scientific discovery process and the evolution of our cosmic understanding. The Hubble constant, H, has been rigorously studied and refined for decades. Different measurement techniques and data sets have led to the "Hubble tension," with current measurements from the Cosmic Microwave Background (CMB), Cepheids, and supernovae converging on values between 67-73 kilometers per second per megaparsec. This ongoing refinement is essential for accurately dating the universe and understanding its expansion history; a higher H implies a faster expansion and younger universe, while a lower value suggests slower expansion and an older universe.

The redshift-distance relationship discovered by Hubble directly manifests the expansion of spacetime itself, not just galaxies moving through static space. As the "fabric" of the universe stretches, light traveling through it has its wavelength progressively elongated. The longer light travels, the more stretching occurs, explaining why distant galaxies exhibit larger redshifts.

Cosmological redshift, a fundamental concept in modern cosmology, allows astronomers to probe the universe's history and large-scale structure. By measuring the redshift of light from distant objects, we look back in time. Light from a galaxy billions of light-years away shows us that galaxy as it was in the distant past when the universe was younger and smaller, enabling the study of galactic evolution, large-scale structure formation, and cosmic changes over epochs.

The redshift of light from galaxies is primarily due to cosmological redshift, the stretching of light's wavelength as space expands, rather than solely the Doppler effect from their motion through space. While the Doppler effect and cosmological redshift are nearly equivalent for nearby galaxies, cosmological redshift dominates for very distant ones. The 'z' value, which quantifies the amount of stretching, indicates greater distance and an earlier epoch in the universe's history.

The discovery of redshift in extragalactic objects and the formulation of Hubble's Law provided the first strong empirical evidence for a dynamic universe, challenging earlier static cosmic models. This was a philosophical as well as a scientific revolution, transforming cosmology from speculation to a data-driven science. The linear relationship between distance and recessional velocity, mapped by Hubble and Humason, formed the observational bedrock of the Big Bang theory. Redshift is the most direct evidence for the universe's ongoing expansion, unequivocally pointing to a hot, dense beginning—the essence of the Big Bang.

While redshift confirms the universe's expansion, a cornerstone of the Big Bang theory, direct evidence for its primordial fiery state lies in the Cosmic Microwave Background (CMB) radiation. Often called the "afterglow" of creation, the CMB is a compelling fossil from the universe's infancy, permeating every corner of the observable universe.

The theoretical basis for this residual radiation was established decades before its discovery. In the late 1940s, George Gamow, Ralph Alpher, and Robert Herman explored the implications of a hot, dense early universe. They predicted that residual heat from this initial epoch would still be detectable, having cooled over billions of years of expansion to a few degrees above absolute zero, uniformly spread across the sky.

This predicted radiation was a logical consequence of the Big Bang hypothesis.

While theoretically predicted, the Cosmic Microwave Background (CMB) remained elusive until its accidental detection in 1964 by Bell Laboratories radio astronomers Arno Penzias and Robert Wilson. Tasked with measuring faint space signals, their sensitive horn antenna consistently picked up a persistent, low-level background hiss from all directions.

This stubborn, uniform noise confounded them. Despite meticulously checking equipment, recalibrating instruments, and even cleaning out pigeon droppings, the hiss persisted. Simultaneously, a team at Princeton University, led by Robert Dicke and including James Peebles, Peter Roll, and David Wilkinson, was independently developing a radio telescope to search for this very "relic radiation" – the predicted afterglow of the Big Bang.

Upon learning of the Princeton group's work, Penzias and Wilson realized the profound significance of their persistent signal. The "noise" they'd been trying to eliminate was, in fact, the cosmic microwave background radiation. Their findings were published back-to-back in the *Astrophysical Journal* in 1965. Penzias and Wilson's detection of this isotropic microwave radiation, initially measured at approximately 3.5 Kelvin (later refined to 2.7 Kelvin), provided the first strong, direct observational evidence for the Big Bang, confirming the universe's hot, dense early phase.

The CMB's properties align remarkably with the Big Bang model. Its near-perfect uniformity across the sky is a key prediction, indicating a hot, dense, and homogeneous early universe. Penzias and Wilson's discovery confirmed this isotropy. However, perfect uniformity would contradict the large-scale structures we observe today, like galaxies. This is where more refined CMB observations became crucial.

The true power of the CMB as a cosmological probe emerged with highly sensitive satellite experiments. NASA's Cosmic Background Explorer (COBE) satellite, launched in 1989, was the first major mission to map the CMB with unprecedented precision. COBE's revolutionary results confirmed the CMB's near-perfect blackbody spectrum, corresponding to 2.725 Kelvin, strongly supporting its origin from a hot, dense, and uniform early universe.

COBE's most stunning discovery, announced in 1992, was the detection of tiny temperature fluctuations, or anisotropies, in the CMB. These minuscule variations (parts per 100,000) are paramount. They represent the primordial density fluctuations—the "seeds"—from which all cosmic structure, including galaxies, stars, and planets, eventually grew.

Imagine the early universe as a nearly perfect, hot soup. COBE revealed infinitesimally small temperature and density variations within this soup. Slightly denser, cooler regions and slightly less dense, hotter regions imprinted these tiny differences onto the CMB. Over billions of years, gravity amplified these initial overdensities, leading to the formation of the large-scale structures we observe today. The CMB provides a snapshot of the universe approximately 380,000 years after the Big Bang, when the universe cooled enough for protons and electrons to form neutral atoms, making it transparent to photons. This period is known as recombination or, more accurately, decoupling.

Following COBE, the Wilkinson Microwave Anisotropy Probe (WMAP) (2001) and the European Space Agency's Planck satellite (2009) provided even more detailed maps of these CMB anisotropies, measuring tiny temperature differences with extraordinary precision. The characteristic pattern of these hot and cold spots across the sky provides crucial information about the universe's geometry, composition, and age.

By analyzing the statistical properties of these fluctuations—their size, distribution, and amplitude—cosmologists can accurately constrain fundamental cosmic parameters. For instance, the relative sizes of the dominant hot and cold spots in the CMB map are sensitive to spacetime curvature. Data from WMAP and Planck indicate that the universe is spatially flat, meaning Euclidean geometry applies on large scales—a critical prediction of the inflationary model of the early universe.

The universe's composition and early history are revealed through the Cosmic Microwave Background (CMB). By analyzing the CMB power spectrum, scientists have determined that ordinary matter accounts for 5% of the universe's total energy density, dark matter 27%, and dark energy 68%. This provides independent evidence for these components, which are otherwise difficult to detect.

The CMB also offers insights into the universe's initial moments, predating the epoch of decoupling. The observed temperature anisotropies align with predictions from inflationary models, which hypothesize a period of rapid expansion shortly after the Big Bang. This expansion smoothed out initial irregularities and stretched quantum fluctuations, providing the seeds for the large-scale structures seen today.

Essentially, the CMB serves as a cosmological Rosetta Stone, offering a direct view into the early universe and encoding information about its evolution and fate. Its near-uniform temperature indicates the initial homogeneity of the Big Bang, its blackbody spectrum confirms its thermal origin, and its subtle anisotropies reveal details about structure formation, spacetime geometry, and cosmic composition. Ongoing CMB research continually refines our understanding of the Big Bang and the fundamental laws governing the cosmos. It's a "silent, omnipresent echo of creation," reminding us of the universe's fiery beginning and helping us determine its age with remarkable accuracy (around 13.8 billion years). CMB analysis also sheds light on the epoch of reionization, when the first stars and galaxies formed, re-ionizing the neutral hydrogen that filled the universe after recombination. The study of the CMB is an active and evolving field, pushing the boundaries of our knowledge about the universe's past, present, and future.

The journey from the enigmatic singularity to the formation of the first stable atomic nuclei is a story of rapid cosmic evolution. This epoch, lasting mere minutes, was a crucible where fundamental forces differentiated, and matter's building blocks were forged.

Immediately after the Big Bang, the universe was in a state of extreme energy density. As it rapidly expanded and cooled, the four fundamental forces—gravity, electromagnetism, and the strong and weak nuclear forces—began to separate from their unified state. This force separation laid the groundwork for the subsequent emergence of particles and their interactions.

The earliest moments, known as the Planck epoch (up to approximately 10^{-43} seconds), remain shrouded in theoretical mystery due to the breakdown of current physics at these extreme conditions. All four fundamental forces are thought to have been unified then. As the universe expanded and cooled, gravity is theorized to have separated first. Subsequently, in the Grand Unification epoch (from 10^{-43} to 10^{-36} seconds), the strong nuclear force separated from the electroweak force. During this incredibly brief period, the universe was a plasma of exotic elementary particles and antiparticles, constantly being created and annihilated.

The inflationary epoch, occurring roughly between 10^{-36} and 10^{-32} seconds after the Big Bang, is a crucial theoretical addition to the Big Bang model, addressing cosmological puzzles like flatness and horizon problems. This era saw an extraordinarily rapid, exponential expansion of spacetime, driven by a hypothetical "inflaton field." The universe expanded by at least 10^{26} in a fraction of a second, smoothing out irregularities and stretching quantum fluctuations to cosmological scales, providing the seeds for large-scale structures. As inflation ended, the inflaton field's energy density converted into a hot, dense plasma of elementary particles, reheating the universe and setting the stage for Big Bang nucleosynthesis.

Following inflation, the universe was a soup of elementary particles: quarks, leptons (like electrons and neutrinos), and their antiparticles, along with photons and gluons. As the universe continued to expand and cool, quarks began to bind, forming protons and neutrons. This phase transition, driven by the strong nuclear force, occurred when the universe was around 10^{-6} seconds old, at temperatures of approximately 10^{13} Kelvin.

Quarks, the fundamental constituents of protons and neutrons, are held together by the strong nuclear force, mediated by gluons. In the incredibly hot and dense early universe, quarks and gluons existed in a deconfined state called a quark-gluon plasma. As the universe expanded and cooled, the strong force became dominant, causing quarks to bind into hadrons (protons and neutrons being the most stable). This hadronization was a critical step in matter formation. The slight asymmetry between matter and antimatter (one baryon remaining for every billion antibaryons annihilated with baryons) allowed for the existence of our matter-dominated universe.

By the time the universe was about one second old, the temperature had dropped to around 10^{10} Kelvin. The universe was then filled with protons, neutrons, electrons, positrons, neutrinos, antineutrinos, and photons. The immense density made the universe opaque, with photons constantly scattering off charged particles. Neutrinos and antineutrinos, interacting only via the weak nuclear force, began to decouple from the plasma, free-streaming through the universe. This neutrino decoupling is a significant event, as their faint relic signal, the Cosmic Neutrino Background, is a theoretical prediction, though currently undetectable.

Big Bang Nucleosynthesis (BBN) describes the crucial period from roughly one second to three minutes after the Big Bang, during which the universe's first light atomic nuclei were formed. As the universe expanded and cooled, protons and neutrons fused into deuterium, an isotope of hydrogen. Deuterium's low binding energy meant it could only form below a specific temperature threshold, preventing its immediate disintegration by high-energy photons.

Deuterium then served as a foundation for heavier nuclei. It fused with protons or neutrons to create tritium and helium-3, respectively. Subsequently, tritium fused with a proton to form helium-4, the most stable light nucleus. Trace amounts of lithium-7 and beryllium-7 (which decayed into lithium-7) were also produced.

The observed abundances of these light elements—approximately 24-25% helium-4, along with deuterium, helium-3, and lithium-7—align remarkably with BBN predictions based on cosmological models. This agreement is powerful evidence supporting the Big Bang theory. Without BBN's precisely controlled conditions, the universe would be almost entirely hydrogen, lacking the helium and lithium essential for star and heavier element formation.

Nucleosynthesis concluded around 20 minutes after the Big Bang, when the universe's temperature dropped to about 10^8 Kelvin and its density decreased, slowing fusion reactions. The universe at this point was primarily hydrogen (75%), helium (25%), and trace lithium, establishing its primordial chemical composition, which would later be transformed by stellar nucleosynthesis.

Following nucleosynthesis, the universe remained an opaque plasma of nuclei, electrons, and photons for several hundred thousand years, with charged particles scattering light. This state persisted until approximately 380,000 years after the Big Bang, when the universe cooled to about 3000 Kelvin. This epoch, known as recombination or decoupling, saw electrons bind with atomic nuclei to form neutral hydrogen and helium atoms. With free electrons largely bound, photons could travel unimpeded, releasing the Cosmic Microwave Background radiation observed today.

This period, from the Big Bang to the formation of neutral atoms, was a critical foundational stage, setting the fundamental parameters for the universe's composition and structure. The precise ratios of light elements from BBN, alongside initial density fluctuations imprinted on the CMB, are direct legacies of this fiery infancy. Studying these early moments through theoretical models and observational evidence like the CMB and elemental abundances profoundly validates the Big Bang framework, allowing us to trace all matter back to these explosive beginnings.

While the Big Bang model successfully explains many cosmological observations, it still presents enigmas. The journey from the early universe's incandescent plasma to nucleosynthesis and photon decoupling paints a compelling picture of cosmic evolution. However, the model is a work in progress, and its limitations point toward areas where our understanding is incomplete. These unanswered questions are not flaws but rather signposts guiding scientific advancement, encouraging new perspectives to bridge gaps in our knowledge.

The Cosmic Microwave Background (CMB) radiation presents a significant puzzle: the horizon problem. The CMB, a snapshot of the universe at 380,000 years old, displays remarkable uniformity across the sky. Yet, regions separated by more than one degree at that time should have been causally disconnected, meaning light (and thus information) couldn't have traveled between them to establish thermal equilibrium. The observed uniform temperature across these disconnected patches suggests an improbable degree of fine-tuning if the basic Big Bang model is correct.

Connected to this is the flatness problem. Einstein's theory of general relativity describes spacetime geometry by its curvature, which can be positive (eventual collapse), negative (eternal expansion), or flat (eternal, decelerating expansion). CMB observations indicate our universe is remarkably close to spatially flat. This means the universe's density parameter (Ω), the ratio of its actual density to the critical density for a flat universe, is very close to 1. For this to be true, the initial density at the Big Bang must have been incredibly precise. Even a slight deviation would have led to extreme curvature, either rapid collapse or unbounded expansion, long before galaxies and stars formed. This sustained flatness over billions of years implies an almost unfathomable initial precision.

These two problems—the horizon and flatness problems—suggest that the standard Big Bang model alone doesn't fully capture what happened in the universe's earliest moments. This motivated the theory of cosmic inflation, proposed by Alan Guth in the early 1980s and refined by others like Andrei Linde and Paul Steinhardt. Inflation posits an extremely rapid, exponential expansion of spacetime in the first fraction of a second after the Big Bang, expanding the universe by a factor of at least 10^{26}. This massive expansion would have smoothed out initial irregularities, stretching a tiny, causally connected region to encompass the entire observable universe today, thereby solving the horizon problem. Furthermore, inflation would have amplified tiny quantum fluctuations to macroscopic scales, and made an already nearly flat universe even flatter, explaining its observed flatness. These stretched quantum fluctuations are believed to be the "seeds" of the large-scale structures we see today, such as galaxies, clusters, and superclusters.

However, despite elegantly addressing these fundamental issues, inflation introduces new questions. The nature of the "inflaton field," the hypothetical scalar field driving this rapid expansion, remains unknown. Its properties and what triggered its decay (leading to the universe's reheating and the subsequent Big Bang expansion) are still subjects of active research.

Beyond the challenges of the early universe, contemporary astronomical observations reveal even deeper mysteries: dark matter and dark energy. These invisible, undetectable components dominate the universe's mass-energy content, yet their nature remains profoundly unknown.

Evidence for dark matter is multifaceted. Galactic rotation curves show galaxies rotate too fast based on their visible matter, indicating an unseen gravitational source. Similarly, galaxy clusters exhibit higher velocity dispersion than accounted for by visible matter. Gravitational lensing also reveals far more mass than is visible. These observations consistently point to a substantial amount of non-luminous matter that interacts gravitationally but not with light. Leading candidates include weakly interacting massive particles (WIMPs) or axions, hypothetical particles predicted by extensions to the Standard Model of particle physics. However, direct detection of these particles remains unachieved.

The enigma deepens with the discovery of dark energy in the late 1990s. Astronomers found that the universe's expansion is accelerating, not decelerating as expected from gravity. This acceleration implies a repulsive force throughout spacetime, dubbed dark energy. The simplest explanation is the cosmological constant (Λ), a form of energy inherent to spacetime with negative pressure that drives expansion. However, theoretical calculations of vacuum energy density are astronomically larger (by about 120 orders of magnitude) than observed, creating the "cosmological constant problem" and suggesting a fundamental flaw in our understanding of vacuum energy and gravity. Other explanations include quintessence (a dynamic scalar field) or modifications to gravity.

Regardless of the precise explanation, dark energy comprises about 68% of the universe's total mass-energy content, dark matter about 27%, and ordinary matter only about 5%. This means the visible universe—stars, galaxies, and ourselves—is a minor component compared to these unseen entities.

Dark matter and dark energy present considerable challenges to our current understanding of fundamental physics. Their existence and prevalence suggest the presence of physics beyond the Standard Model of particle physics, and potentially a re-evaluation of our current gravitational theories. Unraveling the precise nature of these components and the mechanisms driving cosmic acceleration are among the most urgent questions in modern cosmology and physics. Far from being minor details, they constitute the vast majority of the universe's content and dictate its ultimate fate, yet their fundamental identity remains a profound enigma.

These unanswered questions—encompassing the horizon problem, the flatness problem, the nature of dark matter, and the origin and behavior of dark energy—do not undermine the remarkable success of the Big Bang model. Instead, they act as powerful catalysts for further scientific investigation. They represent the leading edge of our knowledge, compelling us to explore new theoretical frameworks and develop increasingly sophisticated observational techniques. The very incompleteness of the Big Bang model, its ability to point towards phenomena beyond our current explanatory capacity, is precisely what makes it such fertile ground for scientific exploration. Within these very puzzles lies the potential for revolutionary new insights into the origin, evolution, and fundamental nature of our universe. The pursuit of answers to these questions may lead to a revised understanding of gravity, the discovery of new fundamental particles, or even the conceptualization of entirely new paradigms for understanding reality itself.

CHAPTER 5

The Black Hole Big Bang Hypothesis

The Big Bang theory, our standard cosmological model, effectively describes the universe's evolution from a hot, dense state to its current complex form. It successfully accounts for the expansion of space, the abundance of light elements, and the cosmic microwave background radiation. However, this model faces conceptual challenges, failing to explain the cosmic microwave background's uniformity across disconnected regions, the universe's flatness, and the prevalence of dark matter and dark energy. These issues highlight the need for a deeper understanding of the universe's earliest moments and fundamental constituents, leading us to consider alternative hypotheses for cosmic origins.

One such hypothesis, explored in this chapter, proposes that our observable universe might have originated within a black hole. This isn't to suggest the Big Bang was a literal, miniature black hole, but rather that the extreme conditions and unique physics inside a black hole, or a similar process, could have provided the necessary elements for a new universe's creation. The central idea is that matter and energy, compressed to unimaginable densities within a black hole's gravitational field, could undergo a transformative "cosmic rebirth," generating a new, expanding spacetime continuum with the fundamental forces and particles we observe.

To understand this, we must briefly review modern physics' understanding of black holes. A black hole is a region of spacetime where gravity is so intense that nothing, not even light, can escape once it crosses the event horizon. Classical general relativity posits a singularity—a point of infinite density and curvature where known physics breaks down—at its center. However, theoretical frameworks like loop quantum gravity and string theory suggest this singularity might not be infinite, but rather an incredibly compressed, quantum-entangled region where spacetime behaves in ways we are only beginning to comprehend.

Within this context, the black hole Big Bang hypothesis suggests that the formation of a black hole, especially a supermassive one from the collapse of a star or dense matter, could create conditions for a new universe's genesis. The immense gravitational forces would compress matter to densities far beyond any other in the cosmos. This extreme compression could trigger a phase transition or a quantum gravitational event, fundamentally altering spacetime's nature. Imagine the very fabric of reality, under immense pressure, violently rearranging itself or undergoing a quantum tunneling event, effectively "unfolding" into a new, expanding spacetime.

The "Big Bang" of our universe, an initial rapid expansion from a state of extreme density and temperature, could be seen as the "explosion" from a black hole. From the perspective of observers within this new universe, the ejected or reformed matter and energy would appear remarkably similar to the Big Bang, with space, time, matter, and energy systematically evolving. This hypothesis offers a speculative mechanism for seeding a new universe with the fundamental constituents and physical laws of our own.

This framework also offers potential solutions to the "horizon problem" and "flatness problem." If our universe originated within a black hole, the observable universe could have been part of a causally connected, albeit incredibly dense, progenitor system. The subsequent immense expansion would stretch this original region to cosmic scales, explaining the shared thermal history and uniformity of the cosmic microwave background. Similarly, theories of black hole interiors and quantum gravity suggest that extreme densities and pressures can drive spacetime towards a state of minimum curvature, which corresponds to flatness. Thus, a universe unfolding from such a state would naturally be born with a high degree of spatial flatness.

The "parent" universe from which this black hole, and subsequently our universe, emerged, could be part of a cyclical universe or a multiverse. In this view, black holes act as "seeds" for new cosmic epochs, with matter and energy being transmogrified and re-emerged as the building blocks of a new universe. This concept of cosmic recycling suggests an eternal, dynamic cosmos constantly undergoing cycles of birth, death, and rebirth.

Classical physics views singularities as endpoints of spacetime, but quantum gravity theories suggest they might be transitional phases. Instead of a catastrophic collapse, matter and energy might undergo a "bounce" or phase transition under extreme gravitational stress, initiating a new expansion and birthing a new universe. The black hole's event horizon, in this context, could be a cosmic membrane transforming and re-emitting matter and energy between an old and new universe.

The idea of an internal black hole "singularity" as an origin point for a new universe has been explored in various theoretical contexts. It's been proposed that the singularity might be a gateway to another region of spacetime or even another universe. The intense gravitational field could warp spacetime so severely that an apparent collapse for an infalling observer might, from a quantum perspective, lead to a reconnection or a transition to a different state of existence. This extreme compression could force quantum states to reorganize, releasing vast amounts of energy and triggering the creation of new particles and forces in a rapidly expanding volume of space.

A compelling aspect of this hypothesis is its potential to address the origin of dark matter and dark energy. If our universe emerged from a black hole, their peculiar properties might be explained by the specific physical processes that occurred during the black hole's "reorganization" or "bounce." The initial conditions within the black hole environment could have favored the production of exotic particles (dark matter) or a fundamental property of the quantum vacuum (dark energy), driving the observed accelerated expansion. The black hole's concentrated mass-energy could be re-emitted to form the constituents of our universe, including its dark components.

Within a black hole's core, quantum gravitational effects are expected to dominate. Theoretical explorations in loop quantum gravity suggest spacetime has a granular structure at the Planck scale. When matter collapses to Planck density, quantum pressure could resist further collapse, leading to a "Planck bounce." This bounce, an event of immense energy release and rapid expansion, would initiate a new phase of cosmic evolution, effectively replacing the singularity with a quantum bounce, and the resulting expansion would be the Big Bang of our universe.

This quantum bounce scenario offers a powerful alternative to the classical singularity. Instead of an infinitely dense point, we have an extremum of density where quantum gravity effects prevent further collapse. The energy released during this bounce would be the primordial energy of our universe, explaining its extreme density and temperature as required by the Big Bang model. The "event horizon" could be the boundary from which this new universe emerges, shielding internal processes from the outside.

Furthermore, the idea that our universe might be contained within a black hole in a larger, older universe could resolve the issue of initial conditions. Instead of explaining the Big Bang's origin from nothingness, we consider a process of transformation and re-emergence. The properties of the "parent" universe, its fundamental constants, and energy content could influence the characteristics of the "daughter" universe born from its black holes. This could naturally explain the specific values of fundamental constants and particle masses in our universe, perhaps selected from a range of possibilities inherent in the parent cosmos.

The hypothesis that our universe originated from a black hole, while speculative, presents profound observational implications. Evidence of such an origin might be found in subtle imprints on the cosmic microwave background (CMB) or the universe's large-scale structure. Gravitational lensing patterns or specific CMB anisotropies could betray a black hole-like genesis. Detecting these signatures would demand extremely precise measurements and advanced theoretical modeling, but the reward of understanding the universe's ultimate origin makes the effort worthwhile.

This hypothesis also redefines our understanding of time. In the standard Big Bang model, time begins with the singularity. However, if our universe was born from a black hole, time would have pre-existed in a "parent universe," with the Big Bang marking the beginning of time for *our* universe. Processes within the black hole, even if seemingly instantaneous from our perspective, would unfold within the temporal framework of the parent cosmos. This reframes the question of "what happened before the Big Bang?" as an inquiry into the prior state of a larger cosmic environment, rather than an absolute beginning.

Black holes, these cosmic enigmas that warp spacetime, could therefore be seen not as destructive endpoints but as conduits for cosmic creation. Supermassive black holes at galactic centers, and even stellar-mass black holes, could be nurseries for future universes. This paints a picture of an eternally cycling universe, where black holes act as engines of cosmic regeneration. This cyclical model, with universes born from the remnants of previous ones, offers a compelling alternative to a singular, unrepeatable Big Bang.

Furthermore, this hypothesis provides a potential solution to the black hole "information paradox." Instead of information being lost when it falls into a black hole, this model suggests it is transformed and re-encoded into the fundamental laws and particles of a new universe. The quantum state of collapsing matter is preserved in a hidden form, then re-expressed in the expanding spacetime of the "daughter universe," implying a conservation of fundamental information across cosmic generations.

While highly speculative, the idea of our universe originating from a black hole pushes the boundaries of theoretical physics, forcing us to question assumptions about singularities, the beginning of spacetime, and the ultimate fate of matter and energy. Exploring such hypotheses not only seeks to answer fundamental questions about our cosmic origins but also drives the development of new theoretical tools and observational strategies. The universe's complexity may hide secrets stranger than we imagine, and the black hole Big Bang hypothesis offers a tantalizing glimpse into a deeper reality, suggesting that the cosmos's most enigmatic objects might be the cradles of new universes.

The event horizon, traditionally viewed as an inescapable boundary and a cosmic prison, takes on a new meaning within this speculative framework—a threshold to a new beginning. Classical general relativity depicts the event horizon as a one-way membrane, drawing everything towards the singularity. However, as our understanding of gravity at its extremes delves into the quantum realm, this classical picture blurs. Theoretical explorations suggest a radical transformation of spacetime and matter as it approaches and crosses the horizon.

The extreme tidal forces near a black hole, famously causing "spaghettification," demonstrate gravity's differential pull. While these forces tear apart macroscopic objects, this hypothesis leverages their interplay with quantum phenomena.

As intensely compressed and stressed matter nears the event horizon, it experiences exponentially intensifying gravitational gradients. This extreme spacetime curvature is precisely where quantum effects, typically negligible, are predicted to become dominant.

In theories attempting to unify gravity and quantum mechanics, such as loop quantum gravity, a black hole's classical singularity is not an infinitely dense point, but a region of finite, albeit astonishingly high, density and curvature. Spacetime is envisioned with a granular structure at the Planck scale. As matter collapses, a "quantum pressure" is thought to resist further compression, preventing a true singularity. Instead, a "bounce" occurs—a transition from a collapsing to an expanding phase. This quantum bounce, at densities far exceeding nuclear density, is proposed as the crucible for a new universe.

In this context, the event horizon becomes intrinsically linked to this quantum bounce, functioning as a dynamic interface rather than a static barrier. As infalling matter crosses the event horizon, it enters a region where the direction towards the singularity becomes a direction in time, not space. If a quantum bounce intervenes before a classical singularity is reached, the infalling matter would transition from extreme compression to rapid expansion. This transition, triggered by quantum gravity effects near the Planck scale, would manifest as the initial moments of a new Big Bang. Thus, the event horizon could be seen as the boundary separating the final moments of a collapsing phase in a parent universe from the initial moments of an expanding phase in a nascent universe.

The idea of a quantum bounce occurring inside a black hole offers a novel interpretation of the singularity, transforming it from an endpoint of destruction into a nexus of cosmic rebirth. From the perspective of matter falling into a black hole, extreme time dilation means a distant observer sees it freeze at the event horizon, never crossing. However, the infalling matter continues its journey, experiencing a sudden reversal—a transition from extreme density to rapid expansion—if a quantum bounce happens just beyond the event horizon. In this view, the event horizon becomes the fleeting moment of transition, marking the end of the inward journey and the beginning of the outward one.

This scenario redefines the singularity as a point of quantum gravitational activity and cosmic metamorphosis, where extreme density and curvature allow quantum effects to orchestrate a new universe. The event horizon thus functions as both the point of no return for the black hole's previous state and the origin point for a new existence.

Furthermore, the characteristics of the progenitor black hole, such as its mass and angular momentum, could influence the initial conditions of the new universe. For instance, a rotating black hole might impart spin to the nascent universe, potentially explaining cosmic asymmetries or influencing large-scale structure formation. The energy for this new universe would come from the gravitational binding energy of the infalling matter and quantum vacuum energy released during the bounce.

This hypothesis could resolve long-standing cosmological puzzles. The **horizon problem**, which concerns the uniformity of the cosmic microwave background (CMB) across causally disconnected regions, finds an explanation if these regions were once in close proximity within the black hole before the bounce. Subsequent expansion would stretch these adjacent regions to their current vast separation while preserving their initial thermal equilibrium. The event horizon, as the boundary of the black hole interior where the universe originated, would encapsulate what becomes our observable universe, ensuring causal connection for all its points.

Similarly, the **flatness problem**, which highlights the universe's remarkable flatness and the fine-tuning required in the standard Big Bang model, can be addressed. Quantum gravity theories suggest that at extremely high densities, spacetime naturally tends towards minimal curvature (flatness). A quantum bounce at such densities would naturally impart flatness to the nascent universe, preserved and amplified by subsequent expansion. The event horizon delineates the quantum state that establishes this initial flatness.

The **information paradox**, which questions what happens to information falling into a black hole, also gains a new perspective. If a black hole interior leads to a new universe, information is not lost but rather scrambled and re-encoded into the quantum states and fundamental parameters of the new cosmos. The intricate quantum entanglement within the black hole could determine the fundamental laws and constants of our universe. In this context, the event horizon is not an information terminator but a boundary facilitating its transformation.

While the precise physics at the event horizon remains a frontier of theoretical research, models like some string theory scenarios propose that black holes effectively "unfold" matter into new, expanding geometries. In such models, the event horizon is crucial for mediating this transition, acting as an impenetrable boundary from the outside yet an origin point from within.

Another way to visualize this connection is through a conformal transformation. Some theoretical frameworks suggest that a black hole's interior, especially as described by advanced quantum gravity, can be conformally mapped to an expanding de Sitter spacetime—a flat spacetime exhibiting accelerated expansion. This implies that the physics within the black hole's event horizon could be equivalent to the physics of a new, expanding universe. The event horizon, in this analogy, acts as the boundary translating the physics of the "old" collapsing spacetime into the physics of the "new" expanding spacetime.

The implications of linking the event horizon to the Big Bang are profound, suggesting that our universe may not be a singular event but part of a grander, cyclical or emergent cosmic process. Each black hole could serve as a cosmic seed for a new universe, making our universe one of potentially many, born from previous cosmic epochs, with each event horizon acting as a cosmic womb. This cyclical view offers an eternal and dynamic perspective where creation and destruction are integral to a continuous cosmic narrative.

The immense energy required for a Big Bang finds a plausible source in a black hole bounce. The gravitational potential energy of collapsing matter combined with amplified quantum vacuum energy at extreme densities could provide the impetus for rapid expansion. The event horizon, marking where this energy is maximally concentrated and quantum effects dominate, becomes the crucial locus for this energy release.

Understanding the matter and energy of a new universe emerging from a black hole is key. It is theorized that fundamental forces and particles would be dictated by the specific quantum gravitational state achieved during the bounce. This could mean different black holes give rise to universes with slightly varied physical laws or constants, supporting the idea of a multiverse. The event horizon, as the interface, would thus be responsible for "selecting" or "generating" these fundamental properties.

Observational signatures of this hypothesis are subtle and challenging to detect, often at the limits of current technology. However, theoretical physicists are exploring potential imprints, such as specific patterns in large-scale matter distribution, subtle correlations in the cosmic microwave background, or gravitational wave signals from the quantum bounce. The event horizon, as the boundary of this origin, might imprint signatures on the universe's large-scale structure, traceable back to its black hole genesis.

Ultimately, the connection between the event horizon and the Big Bang highlights the power of theoretical physics to unite seemingly disparate phenomena into a cohesive, albeit speculative, narrative. It challenges our classical understanding of black holes as objects of pure destruction, instead positing them as potential cradles of creation. The event horizon, once a symbol of finality, might instead represent the ultimate cosmic boundary, a veil hiding the fiery birth of universes while simultaneously acting as the gateway to their existence. By linking these monumental concepts, this hypothesis invites us to reconsider the fundamental nature of cosmic origins and the profound, possibly cyclical, destiny of spacetime itself.

The term "spaghettification" dramatically describes a cosmic process where matter is stretched and torn apart by the immense gravitational forces near a black hole. This occurs because the gravitational pull on an object's closer parts is significantly stronger than on its farther parts, causing it to elongate into a thin, spaghetti-like strand and compress perpendicularly. This phenomenon is a direct consequence of Einstein's theory of general relativity, which explains how mass warps spacetime. As an object nears the black hole's singularity, where spacetime curvature classically becomes infinitely steep, tidal forces are thought to become infinite.

However, the Black Hole Big Bang Hypothesis offers a revolutionary reinterpretation of spaghettification. Instead of utter destruction, it proposes a profound cosmic metamorphosis. This extreme stretching, particularly under quantum-gravitational conditions, could actually initiate a new universal expansion rather than annihilate matter.

Imagine this not as simple stretching, but as an extreme distortion of spacetime itself. As matter, like a star or gas cloud, crosses the event horizon and spirals inward, spacetime fabric is intensely stretched and warped. While the classical singularity is a point of infinite density, in the realm of quantum gravity, it's replaced by a region where spacetime becomes quantized and granular, resisting infinite compression. Within this quantum realm, spaghettification gains a new role. The intense stretching could "unfold" or "disperse" matter across a rapidly expanding spacetime, effectively carrying the stretched matter into a new, expanding state rather than a single point.

This reinterpretation hinges on the behavior of matter and spacetime at the Planck scale, the smallest theoretical scale where quantum gravity dominates. Theories like loop quantum gravity suggest spacetime isn't continuous but composed of discrete "loops" or "grains." As matter collapses into a black hole, it's compressed into these quantized units. The immense pressure at these densities triggers a "quantum bounce"—a reversal from collapse to outward expansion. In this context, spaghettification accelerates matter and energy to these critical Planck densities, preparing them for this quantum bounce. The extreme tidal forces, by aligning and stretching matter radially, efficiently concentrate spacetime and matter components for this fundamental transition.

As infalling matter undergoes spaghettification, its constituent particles aren't just ripped apart; their relative positions within spacetime are drastically altered. The stretching of spacetime itself dramatically increases the radial distances between these particles. Crucially, within a quantum-gravitational framework, these stretched distances don't lead to an infinitely small point. Instead, they extend across a dynamically evolving, expanding spacetime. The once-bound matter, stretched into a strand, then finds itself stretched outward as the universe inside the black hole begins its expansion. Thus, the spaghettified form is a transient, extreme phase before the material disperses and reconfigures as the initial conditions of a new universe.

The energy fueling this process is enormous. The gravitational potential energy of the black hole's matter, along with amplified quantum vacuum energy at extreme densities, powers this new Big Bang. Spaghettification, by pulling matter and energy deep into the black hole's gravitational well, concentrates this potential energy to the point of release in the most energetic event imaginable: the birth of a new cosmos. The destructive stretching force also accelerates spacetime components towards a critical density. It's like stretching a rubber band to its breaking point; instead of snapping, it recoils and expands with tremendous force. Here, the black hole's interior stretches spacetime and matter, and the quantum bounce is the recoil initiating a new, expansive phase.

The implications for cosmic structure formation are also fascinating. The way matter is stretched and dispersed during spaghettification could influence the initial distribution of matter and energy in the nascent universe. Any inherent angular momentum or internal structure in the infalling matter might be imprinted onto the expanding spacetime. This could potentially explain observed large-scale anisotropies in the cosmic microwave background or the distribution of galaxies and clusters in our own universe. The specific pattern of stretching during spaghettification could provide the initial "seeds" for the cosmic web, for instance, leading to filamentary structures if matter streams into the black hole in a particular configuration.

Spaghettification, often considered a destructive process, can be reinterpreted through the lens of holographic principles and conformal field theories as a mechanism of cosmic rebirth. In some theoretical models, the extreme stretching of spacetime during spaghettification might be a physical manifestation of information "unfolding" onto an expanding boundary of a new universe. This suggests the event horizon is not just a point of no return but a dynamic interface for cosmic transformation.

Shifting to a quantum perspective, spaghettification's extreme tidal forces, which stretch matter, are fundamentally linked to spacetime's curvature. When this curvature reaches a critical point, as predicted by quantum gravity near the Planck scale, spacetime itself transitions. The stretching doesn't just tear objects apart; it stretches the fabric of spacetime until it can no longer sustain a collapsing structure, instead rebounding into an expanding state, carrying the matter with it.

This reinterpretation transforms spaghettification from ultimate destruction into a harbinger of creation. The very forces perceived as inimical to existence become architects of new existence, pulling apart the threads of spacetime to create conditions for a fresh start. The once-feared tidal forces are seen as "midwives" of a new universe, preparing and dispersing matter and energy for cosmic genesis. This powerful reframing illustrates how, at extreme scales, annihilation can be a prelude to a grander reality, repurposing a terrifying phenomenon into a fundamental mechanism for cosmic rebirth.

The Black Hole Big Bang Hypothesis intertwines with cyclic cosmology and the "Big Bounce" concept, suggesting the universe is a continuous, pulsating entity. Within this framework, the catastrophic collapse into a black hole, seemingly destructive, is re-imagined as the precursor to a new cosmic epoch. This aligns with Big Bounce scenarios, where the universe, instead of ending in a Big Crunch, rebounds into a new phase of expansion.

Traditional Big Crunch models depict the universe contracting after expansion, culminating in an infinitely dense singularity. However, the Big Bounce modifies this by proposing that at extremely high densities, quantum gravitational effects prevent a true singularity, instead providing repulsive pressure that "bounces" the universe back into an expansive phase. Our black hole scenario provides a potential location for this mechanism. The immense gravitational pull within a black hole, compressing matter to extreme densities, mirrors the contracting phase of a cyclic universe, potentially triggering a Big Bounce.

The connection between the black hole singularity and the Big Bounce is rooted in the interplay of extreme gravity and quantum mechanics. In loop quantum gravity, spacetime is quantized. As matter collapses into a black hole's center, it compresses into these discrete units. At a certain density, the quantum geometry of spacetime resists further compression, manifesting as a powerful repulsive force that initiates expansion—the essence of the Big Bounce. Thus, the interior of a black hole, where matter reaches ultra-high densities, could be the crucible for a Big Bounce, birthing a new, expanding universe. Infalling matter, subjected to spaghettification and extreme compression, is not destroyed but reconstituted and flung outwards as nascent matter for a new cosmos.

This cyclical view offers a compelling answer to "what came before the Big Bang." If our universe began with a Big Bang preceded by a Big Crunch, the question merely shifts. By positing the black hole interior as a site of cosmic genesis, we suggest each black hole could be a "mother" universe for a new cosmos. This implies a universe teeming with black holes, each a seed for a future universe. Our own Big Bang, in this grander scheme, could have originated from matter collapsing within a black hole in a prior cosmic epoch. This perspective transforms the solitary, cataclysmic Big Bang into one event in an endless series of cosmic rebirths, each born from a collapsed predecessor.

The interior of a black hole's singularity, when viewed through the lens of quantum gravity, remarkably mirrors the conditions necessary for a Big Bounce. While classical general relativity predicts an infinitely dense point, quantum mechanics intervenes at the Planck scale. Theories like loop quantum cosmology (LQC) resolve the singularity, replacing it with a region of maximal, yet finite, density and curvature. At this peak density, the quantum geometry of spacetime transitions, repelling itself and initiating an expansion, effectively replicating Big Bounce conditions due to immense gravitational forces inside a black hole. Matter falling into a black hole isn't just accumulating; it's being compressed to densities that could theoretically rival or exceed those of the Planck era. Extreme tidal forces, spaghettification, and eventual compression all contribute to driving matter and energy toward this quantum-gravitational turning point.

Imagine a black hole's interior as a cosmic incubator. Matter and energy drawn from the parent universe are subjected to unimaginable forces, stretched, compressed, and ultimately concentrated into an incredibly dense, transitional state. At this critical juncture, quantum spacetime asserts itself, with repulsive forces of quantum gravity overcoming inward gravitational pull, causing a "bounce"—the birth of a new universe. Matter and energy, once seemingly destined for destruction, are reconfigured and propelled outwards into a new spacetime, initiating a new Big Bang. This new universe then expands, forming new black holes and perpetuating the cosmic cycle.

The concept of a cyclical universe, or one born from a Big Bounce, is an area of intense research. Cosmologists search for subtle signatures in the cosmic microwave background (CMB) radiation, the afterglow of our own Big Bang, that might hint at a prior cosmic era. For example, some theories suggest gravitational waves from a pre-bounce universe could have imprinted patterns onto the CMB. The precise distribution of matter and energy in the early universe—its observed homogeneity and isotropy—could also be a legacy of the bouncing mechanism. If our Big Bang was a bounce from a preceding contraction, the initial conditions would have been shaped by that prior phase. The Big Bounce hypothesis thus provides a natural framework for understanding these observed cosmological features without relying on speculative theories of cosmic inflation or fine-tuning.

Furthermore, the Black Hole Big Bang Hypothesis offers a unique perspective on the origin of dark energy and dark matter. If our universe emerged from a black hole, the initial conditions would have differed dramatically from those predicted by a standard Big Bang model. The intense energy release during the bounce could have produced not only ordinary matter and radiation but also the exotic forms we attribute to dark matter and dark energy. The specific properties of the parent universe, and how its matter and energy were compressed and released, could directly influence the primordial abundance of these components in the new, expanding universe. This offers a potential avenue for testing the hypothesis, as different black hole interior models might predict varying ratios of dark energy to dark matter.

The concept of nested universes, where one exists within another, naturally fits this framework. If black holes are gateways to new universes, then our universe could be one bubble in a vast, interconnected cosmic foam. Each black hole within our universe could, in turn, be an entrance to yet another universe, creating a fractal-like structure of cosmic existence. This grand tapestry of nested realities challenges our conventional understanding of cosmic boundaries and origins, suggesting a far more complex and interconnected reality. The implications are profound: our entire observable universe could be the interior of a colossal black hole within a larger, overarching cosmic structure.

This hypothesis beautifully reconciles seemingly disparate cosmological phenomena. The extreme conditions inside a black hole, the universally accepted Big Bang model, and the theoretical framework of cyclic cosmologies and the Big Bounce all converge. Gravity's destructive force, which forms black holes, is reframed as a generative force for new universes. The singularity, once an enigma of infinite density, becomes a transitional point of quantum rebound. This perspective offers a more elegant and unified picture of cosmic evolution, suggesting an unbroken chain of creation and destruction, expansion and contraction, all intricately linked through the enigmatic nature of black holes.

The crucial question is how matter gets "prepared" for a bounce within a black hole. Spaghettification plays a vital role, not just stretching matter into oblivion, but aligning and dispersing it across rapidly changing spacetime. In a bounce context, this preparation might involve orienting quantum vacuum fluctuations or the fundamental constituents of spacetime itself in a way that favors outward expansion. The extreme tidal forces could, in essence, "tune" the quantum field to a state conducive to a new inflationary epoch, like winding up a cosmic spring. Infalling matter and black hole gravitational forces are the winding mechanism; the quantum bounce is the release, propelling the universe outwards.

Moreover, the black hole's mass and type could influence the resulting universe's characteristics. A supermassive black hole, formed from a collapsing star or galaxy, might produce a universe with different initial conditions than a stellar-mass black hole. The initial distribution of matter, angular momentum, and energy within the collapsing object could be imprinted onto the subsequent expansion. This offers a fascinating possibility: that our universe's properties—its expansion rate, matter content, and large-scale structure—could be directly influenced by the properties of its originating black hole. This hypothesis introduces a level of cosmic genealogy, where the "parent" black hole shapes its "offspring" universe.

Observational signatures of such a scenario are subtle but potentially detectable. If our universe was born from a black hole bounce, there might be subtle correlations between the properties of existing black holes in our universe and its large-scale structure or cosmic parameters. For instance, a particularly massive or rapidly spinning black hole in the prior universe that triggered our Big Bang could manifest as a preferred directionality or anisotropy in our current cosmic expansion. While current data strongly supports a highly isotropic and homogeneous universe, future, more precise measurements might reveal such subtle hints of a black hole origin. The search for these signatures is at the forefront of modern cosmology.

The Black Hole Big Bang Hypothesis offers a compelling, though speculative, narrative for cosmic origins. It proposes a universe in a continuous cycle of creation and rebirth, with black holes acting as cosmic "wombs" for new realities. This challenges our current understanding of cosmic finality, opening new avenues for exploring spacetime, gravity, and existence itself. It depicts a dynamic, interconnected, and eternally renewed universe, a testament to the profound and often counter-intuitive workings of the cosmos. Future theoretical advancements and observations will be crucial for validating this hypothesis.

Despite its radical nature, the Black Hole Big Bang Hypothesis finds tentative support in several cutting-edge areas of theoretical physics. These abstract and developing frameworks offer conceptual bases for how extreme conditions within a black hole could trigger a new cosmic genesis, akin to a Big Bounce. Therefore, our exploration must delve into these speculative realms, seeking inspiration and validation from the forefront of our understanding of gravity and quantum mechanics.

One such area is quantum gravity, which aims to unify Einstein's general relativity with quantum mechanics. The classical description of a black hole's singularity, a point of infinite density where physics breaks down, is a major problem. Quantum gravity theories attempt to resolve this by proposing that at the Planck scale (the smallest possible scale), spacetime and matter behave differently. Loop Quantum Gravity (LQG) and String Theory offer distinct approaches.

In LQG, spacetime is granular, made of discrete "loops." As matter collapses into a black hole, it's compressed into these fundamental units. LQG predicts that at extreme density, the quantized nature of spacetime creates a repulsive force, preventing an infinite singularity. Instead, it suggests a "bounce" – a transition from contraction to expansion. This quantum bounce is central to our hypothesis. The intense gravitational pressure, which would lead to a singularity in classical physics, is reinterpreted in LQG as the compression of spacetime to its smallest constituents, leading to a rebound. The process of matter falling into a black hole, subjected to increasing densities, could drive spacetime to this quantum-repulsive threshold. Infalling matter isn't annihilated but becomes part of the quantum spacetime fabric, then ejected in a new cosmic expansion.

String Theory presents a different view, positing that fundamental constituents are tiny, vibrating strings. Some String Theory models, especially those with extra spatial dimensions and branes (higher-dimensional objects), can also resolve singularities. For example, certain brane-world scenarios suggest that gravity might leak into these extra dimensions, altering its behavior at small scales and high densities. It's conceivable that the extreme curvature and density within a black hole could lead to such a scenario, where extra dimensions become significant, facilitating a collapse-to-expansion transition. The "collision" of branes or the dynamic evolution of spacetime geometry in these models might offer mechanisms analogous to the quantum bounce in LQG. The immense energy density inside a black hole could, in this context, trigger a phase transition in the extra-dimensional geometry, leading to a rebound.

Beyond these specific quantum gravity frameworks, broader theoretical considerations support a cyclical or bouncing universe, which our hypothesis leverages. Emergent gravity, for instance, suggests that gravity might not be a fundamental force but an emergent phenomenon from more basic degrees of freedom, perhaps related to quantum entanglement or spacetime thermodynamics. If gravity is emergent, its behavior at extreme conditions, like those in black holes, could be radically different, potentially allowing for repulsive effects leading to a bounce.

Furthermore, the nature of the vacuum state and quantum fluctuations is crucial. In a Big Bounce context, the quantum vacuum state before the bounce is paramount. It's theorized that during the contracting phase, these quantum fluctuations can grow and evolve, eventually dominating and driving expansion. The extreme conditions within a black hole might provide a unique environment for these fluctuations to amplify, preparing the quantum vacuum for a new epoch of expansion. Imagine the intense gravitational field as a powerful amplifier for pre-existing quantum instabilities. As matter is compressed, it could effectively "energize" these fluctuations to a critical level, triggering the rebound.

The concept of "cosmic forgetfulness" is also relevant. Bouncing cosmologies often address how information and entropy from a previous cosmic cycle are "reset" or transformed during the bounce. If our universe originated from a black hole's interior, the immense compression and transformation of matter and energy could naturally act as this reset mechanism. Extreme tidal forces and gravitational gradients might thoroughly mix and scramble information, providing a clean slate for the new universe. This aligns with the idea that the singularity isn't a point of information loss (as in the classical black hole paradox) but a point of extreme transformation and re-initialization. Infalling matter isn't destroyed but reconfigured at a fundamental level, its prior history potentially obscured by the sheer intensity of the transitional process.

The mathematical formalism for studying cosmological singularities in general relativity and their resolution in quantum gravity also offers insights. While classical general relativity breaks down at the singularity, equations describing the approach to it—especially in anisotropic and inhomogeneous cosmologies—reveal that the singularity's nature can depend on initial conditions and system symmetries. In a black hole interior, the collapse is highly anisotropic and influenced by the black hole's mass and spin. This complexity might provide avenues for singularity resolution that differ from a simple, uniform Big Crunch. Such anisotropic resolutions could naturally lead to anisotropic bounces, imprinting specific large-scale features onto the nascent universe.

Our hypothesis finds further theoretical backing in modified gravity theories. While general relativity has been incredibly successful, it's possible that at extremely high densities or curvatures, Einstein's equations might not fully apply. Some modified gravity theories suggest repulsive gravitational effects under specific conditions, which could manifest as a "bounce." If conditions within a black hole trigger these gravitational modifications, a bounce becomes a natural outcome. This could involve adjustments to the Einstein-Hilbert action or the introduction of new fields that mediate gravity unusually at extreme scales.

The enigmatic nature of dark energy also points to possibilities beyond the standard cosmological model. Some theories propose that dark energy is connected to vacuum energy, quantum fluctuations, or the dynamic properties of spacetime itself. If our universe originated from a black hole bounce, the specific quantum gravitational processes involved in that bounce could directly influence the initial abundance and behavior of dark energy. This might explain the universe's observed accelerated expansion without requiring a cosmological constant or quintessence as fundamental entities. The energy density released during the bounce itself could possess properties that mimic dark energy in the early stages of the new universe.

Furthermore, theories exploring the fundamental structure of spacetime, such as causal set theory (which posits spacetime as a discrete structure ordered by causal relations), offer an alternative perspective on singularities. In such discrete spacetimes, infinite densities might be ill-defined. The approach to a singularity could naturally lead to a change in the causal structure, potentially initiating an outward propagation resembling a bounce. The intense compression within a black hole would represent a point where the discrete elements of spacetime are packed so densely that their causal relationships fundamentally change, leading to a new ordering and expansion.

The theoretical landscape is vast and often speculative, yet these diverse research avenues converge on a crucial point: the classical breakdown at singularities, whether in cosmology or black holes, is likely an artifact of incomplete theories. Quantum mechanics, extra dimensions, emergent phenomena, and discrete spacetime structures all provide mechanisms to bypass or resolve these singularities, often leading to a "bounce" or a transition rather than an absolute end. Our hypothesis places the black hole, a well-established astrophysical object, as the site where these quantum gravity effects could manifest most dramatically, bridging the gap between the micro-physical and the macro-cosmological.

Rigorous study of these theoretical frameworks is essential for assessing the plausibility of the Black Hole Big Bang Hypothesis. It underscores that while the idea may seem outlandish, it is an exploration driven by the very frontiers of physics, seeking to understand the deepest mysteries of the cosmos. The quest to reconcile gravity with quantum mechanics is not merely an academic exercise; it is the very foundation upon which such transformative cosmological ideas can be built. The elegance of a universe born from the remnants of its predecessor, encapsulated within the gravitational well of a black hole, is a testament to the power of theoretical physics to reshape our understanding of reality.

CHAPTER 6

Observational Clues and Indirect Evidence

The Cosmic Microwave Background (CMB) is a cornerstone of the Big Bang model, representing the universe's afterglow approximately 380,000 years after its inception. This "snapshot" reveals a universe rapidly expanding and cooling from an extremely hot, dense early state, becoming transparent to light for the first time. Its remarkable uniformity is punctuated by tiny temperature fluctuations, known as anisotropies. These anisotropies seeded the large-scale structures we see today and have been meticulously mapped by missions such as COBE, WMAP, and Planck. The information imprinted on these primordial photons offers profound insights into the universe's composition, geometry, and evolution.

However, as our observational capabilities and theoretical understanding advance, subtle deviations and unexpected features within the CMB continue to emerge, fueling scientific debate and new research avenues. Our hypothesis, which proposes that the universe originated from the explosive dispersal of matter and energy from within a black hole's event horizon, provides a unique framework for re-examining these observed CMB features. If our universe indeed resulted from such a violent, yet fundamental, event, the CMB should bear the distinct marks of its extraordinary genesis.

Within this "Black Hole Big Bang" framework, the primordial plasma that released CMB photons would not have been the uniformly isotropic and homogeneous soup typically envisioned. Instead, it would have been a highly energetic, potentially anisotropic outflow originating from a specific, incredibly dense, and energetic point in spacetime. The inherent nature of a black hole's interior and the process by which matter and energy might escape, or trigger a new expansion, could imbue the universe's initial conditions with characteristics that diverge from standard inflationary cosmology.

Consider this scenario: matter and energy compressed to extreme densities within a black hole might not explode outwards in a perfectly uniform manner. The immense gravitational forces and the quantum gravity processes driving the "bounce" could impart directional or intrinsic angular momentum to the ejected material. This initial anisotropic expulsion could, in principle, lead to subtle, large-scale patterns in the CMB that are not easily explained by standard models.

One area of particular interest lies in the large-angle anomalies observed in the CMB. While small-scale anisotropies align remarkably well with the Lambda-CDM model, certain features on larger angular scales remain enigmatic. These include the unusual alignment of the quadrupole and octopole moments (the lowest-order spherical harmonics describing temperature variations), apparent parity violation (an asymmetry between opposite hemispheres), and the "cold spot"—a region of the sky with a significantly lower temperature than expected. In the standard cosmological picture, these are often attributed to statistical flukes due to having only one observable universe, or as hints of physics beyond the standard inflationary model, such as specific inflationary potentials or topological defects.

However, our Black Hole Big Bang hypothesis offers a more direct and fundamental explanation for these anomalies. If the universe emerged from a black hole, the dispersal mechanism would likely be highly anisotropic. Imagine the black hole's interior as a complex, quantum gravitational environment. During the "bounce," the ejection of energy and matter might not be spherically symmetric. This process could resemble a highly energetic jet or a directional burst, influenced by the mass, spin, and possibly quantum properties of the parent black hole. Such an anisotropic outflow would naturally imprint a preferred direction or alignment onto the nascent universe. This inherent anisotropy could manifest as the observed alignment of the low-l multipoles in the CMB, which appear to point to a particular axis in the sky, contrary to the expectation of perfect statistical isotropy from inflation. The observed quadrupole and octopole, for example, could be a direct consequence of the preferred direction of the initial burst from the black hole's event horizon.

The "cold spot" is another intriguing anomaly, a vast, cool region in the Eridanus constellation that is statistically improbable within the standard inflationary paradigm. Proposed explanations include its identification as a supervoid (a region with a deficit of galaxies), a consequence of exotic physics like a collision with another bubble universe in a multiverse scenario, or even an indication that our universe is not spatially flat. If our universe was born from a black hole, the initial energy expulsion might have been less uniform than assumed. Perhaps a region of slightly lower energy density or more rapid expansion in a particular direction occurred, leading to a cooler region in the CMB. This could be a direct imprint of the black hole's internal dynamics during the bounce. The spacetime structure inside the black hole, prior to the critical bounce point, could have been such that the outward ejection was not perfectly uniform, resulting in an observed deficit of energy density in one particular angular direction in the CMB.

The hypothesis of matter dispersing from an event horizon offers new insights into the universe's initial conditions. The intense density and energy within a black hole could produce unique primordial gravitational waves or exotic particles not predicted by standard inflationary models. These could leave distinct signatures on the Cosmic Microwave Background (CMB). For example, if quantum gravitational processes during a "bounce" generate a specific spectrum of primordial gravitational waves, these waves would interact with the photon-baryon plasma, imprinting characteristic patterns on CMB anisotropies. The precise shape and amplitude of these patterns would depend on the specifics of the black hole bounce mechanism.

The nature of matter and energy dispersed from the event horizon is also critical. Our hypothesis suggests that infalling matter is transformed, not destroyed, by immense gravitational and quantum forces, potentially creating new fundamental particles or states of matter unlike those typically assumed in the early universe. This could have altered the primordial plasma's constituents and interactions, influencing recombination—the process where electrons and protons formed neutral atoms, allowing photons to travel freely. If the early universe's opacity or acoustic oscillations were affected by these exotic initial conditions, it would leave a traceable imprint on the CMB power spectrum.

Furthermore, the event horizon introduces a boundary condition absent in the standard Big Bang model. While the standard model describes an unbounded expansion from an initial singularity, our hypothesis posits an origin from a bounded spacetime region. The "breaking out" of an event horizon in the Black Hole Big Bang might inject specific energy or angular momentum, leading to a net rotation of the universe or subtle, detectable correlations between temperature and polarization in the CMB. CMB polarization is particularly sensitive to primordial gravitational waves and magnetic fields, and a universe born from a black hole might have unique initial conditions that generate these phenomena.

Planck satellite data has revealed subtle CMB patterns, like hemispheric asymmetries and temperature-polarization correlations, which have undergone intense scrutiny. While some could be statistical fluctuations, their persistence in high-precision data allows for alternative explanations. Our Black Hole Big Bang hypothesis offers a potential framework for understanding these. For instance, if the parent black hole rotated, the outward expulsion of matter and energy might have a directional component, causing slight asymmetry in CMB temperature or polarization distribution, potentially explaining observed hemispheric differences.

Another consideration is the early universe's entropy. The standard Big Bang model, with inflation, explains a high-entropy, nearly isotropic universe from a potentially low-entropy initial state. In the Black Hole Big Bang scenario, the extremely compressed state within the black hole is high in density but not necessarily high in entropy in the conventional sense. Entropy generation would occur during the bounce. The efficiency and nature of this entropy generation during black hole dispersal could yield a specific early universe entropy, influencing CMB details. Significant dissipation or mixing could naturally produce the high entropy and isotropy we observe in the CMB, but with subtle signatures of its more complex origin.

The CMB's approximate 2.7 Kelvin temperature and blackbody spectrum are strong evidence for a hot, dense early universe that expanded and cooled. Our hypothesis does not contradict this; the immense energy released during black hole dispersal would create a plasma that cools to form the CMB. The question is whether the initial energy and matter distribution, and the specific release processes, leave distinguishable imprints from a simple, homogeneous expansion from a point. The CMB's spectral purity is a stringent test. However, if the dispersal involved a rapid, energetic phase transition or quantum gravitational event, it might create a state that, upon expansion, still approximates a blackbody spectrum, but with subtle deviations or anisotropies.

Moreover, precise measurements of CMB power spectrum acoustic peaks provide information on the universe's baryon, dark matter, and dark energy densities. If initial conditions from a black hole differed in these component abundances, or if new exotic particles were generated during the bounce, it could subtly alter peak positions and heights. For example, if the bounce injected significant primordial magnetic fields, these could influence plasma physics and leave a CMB imprint, especially on its polarization. The current concordance model fits these peaks remarkably well, so any new explanation must account for this success while explaining anomalies.

The Black Hole Big Bang hypothesis offers a potential alternative for the initial conditions leading to the observed CMB. Instead of an infinitely dense and smooth singularity, we propose a black hole singularity—a quantum-gravitational object with specific internal dynamics. The "surface of last scattering" would not have been a uniform spherical shell, but rather a complex, possibly turbulent, outflow from the black hole's event horizon. This explosive dispersal would have set the photon-baryon fluid in motion, carrying the imprints of the anisotropic processes of its creation.

The precise modeling of the dispersal process within a quantum gravity framework, and the subsequent calculation of its specific Cosmic Microwave Background (CMB) signatures, presents a significant challenge. This endeavor demands a deeper comprehension of quantum gravity, particularly regarding singularities, event horizons, and the transition from collapse to expansion. Nevertheless, the ongoing effort to understand CMB anomalies serves as a compelling impetus to explore such alternative cosmological scenarios. If the puzzling features of the CMB are genuine deviations from the simplest inflationary models, rather than statistical flukes, then the Black Hole Big Bang model offers a rich source of potential explanations rooted in the extreme physics of black holes and the speculative, yet exciting, frontiers of quantum gravity. The very fabric of spacetime, compressed and warped within a black hole, could have unfurled to create the structured, isotropic, yet subtly anomalous universe observed through the CMB. In this view, the CMB is not merely the echo of a simple bang, but potentially a complex symphony played by the remnants of a collapsed star, offering clues to a cosmic origin far more intricate than previously imagined.

While the prevailing cosmological model, Lambda-CDM, successfully describes numerous astronomical observations—from galaxy distribution to CMB fluctuations—its triumph is tempered by the profound mysteries of dark matter and dark energy. These invisible components constitute approximately 95% of the universe's total mass-energy density and remain stubbornly elusive. Dark matter, inferred from its gravitational influence, provides the scaffolding for cosmic structure, holding galaxies together and dictating the large-scale arrangement of the cosmos. Dark energy, conversely, drives the observed accelerated expansion of the universe. Although the Lambda-CDM model parameterizes their existence and effects, it offers no fundamental explanation for their nature or origin.

Within the framework of a universe born from the explosive dispersal of matter and energy from within a black hole's event horizon, these pervasive enigmas might find entirely novel interpretations. This radical departure from a singularity-based Big Bang model naturally opens avenues for exploring alternative origins for the universe's fundamental constituents and governing forces. If the cosmos is the energetic fallout of a collapsed stellar object undergoing a quantum gravitational transformation, then the properties of that progenitor black hole and the specific processes of its transition to an expanding universe could directly influence the residual components left behind.

Consider dark matter. Its gravitational signature is ubiquitous, yet its lack of electromagnetic interaction renders it invisible to telescopes. Numerous candidates have been proposed, such as weakly interacting massive particles (WIMPs) and axions, all posited as new fundamental particles beyond the Standard Model. However, in the Black Hole Big Bang scenario, dark matter might not be a collection of exotic particles created during a separate phase of cosmic evolution, but rather a direct consequence of the extreme conditions within the black hole itself. The immense densities and exotic quantum states of matter and spacetime near a singularity or an event horizon might naturally lead to the formation of stable, gravitationally interacting entities that do not couple to the electromagnetic force.

The process of matter compression and subsequent dispersal from a black hole could involve states of matter or energy fundamentally different from those encountered in everyday physics. For instance, the quantum gravitational phase transition facilitating the universe's re-expansion might involve the conversion of baryonic matter, or even the vacuum energy itself, into a form that exhibits only gravitational interactions. The intense gravitational tidal forces and resulting quantum fluctuations within the black hole's interior could fragment ordinary matter or warp spacetime in such a way that stable, non-luminous clumps or fields are generated.

These residual entities, persisting through the initial expansion and cooling, would then manifest as dark matter, dictating the gravitational dynamics of galaxies and clusters.

Furthermore, the very nature of the "bounce" from a black hole could imbue the ejected material with specific properties. If the black hole possessed a net angular momentum, for example, the dispersal process might not be perfectly uniform. This could lead to a dark matter distribution that is not smoothly homogeneous but exhibits certain clustering patterns or directional dependencies from the outset. While current observations suggest dark matter is broadly distributed in halos, a more detailed analysis of its large-scale structure might reveal subtle imprints of its anisotropic origin, aligning with potential large-angle anomalies observed in the CMB. This hypothesis suggests that dark matter might not be merely a passive component, but a legacy of the dynamic, violent birth of the universe from a compact, rotating object.

The possibility of dark matter arising from the specific conditions of a black hole singularity also offers explanations for some of dark matter's observed properties. For instance, the cusp-core problem, where simulations predict denser central cusps than observed in some galaxies, could potentially be resolved if the initial dark matter distribution generated by the black hole bounce was already more "cored" than predicted by standard particle dark matter models. Alternatively, self-interaction mechanisms of dark matter, or interactions with residual exotic fields from the bounce, might naturally smooth out these central densities—a phenomenon that particle dark matter models struggle to explain without fine-tuning.

The origin of dark energy presents an even greater conceptual challenge, as it appears to be a property of spacetime itself, driving an accelerating expansion. In the standard model, dark energy is often associated with a cosmological constant (Lambda), representing the energy density of the vacuum, or a dynamic scalar field (quintessence). While these explanations are phenomenologically successful, they lack deep theoretical grounding. The Black Hole Big Bang hypothesis offers a radical alternative: dark energy might be a residual effect of the black hole's spacetime geometry as it transitioned into our expanding universe.

Consider the nature of spacetime at the Planck scale, particularly near a singularity or event horizon. Here, our current understanding of physics breaks down, and quantum gravity effects are expected to dominate. It is conceivable that the very process of a black hole "unraveling" or transitioning into a new, expanding spacetime could leave behind a persistent energetic imprint that behaves like dark energy. This imprint might not be a localized substance or field, but rather an intrinsic property of the emergent spacetime, a kind of "tension" or "elasticity" inherited from the extreme curvature of the progenitor black hole.

One way to conceptualize this is to consider the event horizon as a boundary. As the universe emerges, it carries with it the memory of this boundary condition. The expansive force associated with dark energy could be a lingering consequence of the immense gravitational potential energy contained within the black hole. As the universe expands, this stored potential energy is released in a manner that drives further expansion, analogous to how a stretched rubber band stores potential energy that is released when it snaps back, but in this cosmic context, the "snap-back" is an outward expansion.

Alternatively, the quantum fluctuations and exotic states of matter and energy present within the black hole could, upon dispersal, create a non-zero vacuum energy density that persists throughout cosmic history. Unlike the hypothetical vacuum energy of quantum field theory, which is vastly larger than observed dark energy, this "primordial vacuum energy" would be a specific value dictated by the physics of the black hole bounce. The extreme conditions might lead to a spontaneous symmetry breaking or a phase transition that locks in a specific, non-zero energy density for the vacuum, which then drives the cosmic acceleration. This would offer a more direct, less speculative origin for the cosmological constant.

The consistency of dark energy across cosmic time, mirroring a cosmological constant, suggests that if it stems from a Black Hole Big Bang, the residual effects of the black hole's spacetime structure are stable. This implies dark energy is an inherent property of the emergent spacetime, established during the black hole's transition into our universe, rather than a fluctuating dynamic field.

Furthermore, the Black Hole Big Bang model offers a new perspective on the poorly understood interaction between dark matter and dark energy. If both originated from the same event—the black hole's dispersal—they might be intrinsically linked. For instance, the distribution of dark matter halos could subtly influence local dark energy density, or vice versa, providing observational tests for this hypothesis. Imagine regions of higher dark matter density, formed by anisotropic dispersal, also having slightly different local dark energy values. These subtle variations, though minor on large scales, could be detected in precision cosmological surveys.

The dominance and unique nature of dark matter and dark energy, distinct from ordinary matter, point to a singular origin story, which a universe born from a black hole provides. Ordinary matter might be a fraction of the total energy that underwent a "gentle" transition during dispersal, while dark matter could be from more extreme fragmentation or quantum binding. Dark energy, then, would be the residual, inherent spacetime property liberated from the black hole's gravitational confinement.

The cosmological constant problem, the vast discrepancy between theoretically predicted vacuum energy and observed dark energy density, could also be resolved in this context. If dark energy is a relic of the black hole's spacetime structure, not the vacuum energy of quantum field theory, this discrepancy is circumvented. Its value would be determined by the specific physics of the black hole bounce, not by cumulative quantum field effects, providing a more robust explanation for its precise value and allowing for cosmic acceleration without fine-tuning.

Additionally, the Black Hole Big Bang hypothesis might clarify the coincidence problem: why dark matter and dark energy densities are comparable today, despite evolving differently. If both derived from the same initial event, their relative abundances at the time of the bounce would be set. Their current comparable magnitudes could directly result from this shared origin, for instance, if the black hole dispersal released energy into distinct channels, one leading to dark matter and another to intrinsic spacetime energy.

Exploring these connections demands a sophisticated understanding of quantum gravity and black hole physics. The specifics of a black hole transitioning into an expanding universe remain highly speculative, contingent on the theoretical framework. However, the observational puzzles of dark matter and dark energy strongly motivate exploring such avenues. If these enigmatic components are indeed relics of a Black Hole Big Bang, their observed properties—densities, distributions, and evolutionary behaviors—could directly probe the physics governing our universe's birth. The study of dark matter and dark energy thus becomes a quest to decipher the ultimate nature of our universe's origin, with these invisible constituents potentially holding the most profound clues to its ancient secret: its birth from a black hole's heart.

The universe's remarkable hospitality to life, known as the "fine-tuning problem," highlights that many fundamental physical constants and parameters are precisely valued. Even slight deviations would render the universe sterile. This exquisite calibration, a persistent challenge in cosmology, ranges from the strength of fundamental forces and elementary particle masses to the cosmological constant, all delicately balanced to allow for stars, galaxies, planets, and life.

Consider the gravitational constant, \$G\$. Stronger gravity would accelerate stellar fuel consumption, preventing planetary formation and life's evolution. Weaker gravity would preclude star and galaxy formation entirely, as matter wouldn't coalesce. The electromagnetic force, governing atomic and molecular structure, is similarly finely tuned. A minuscule deviation in the ratio of electromagnetic to gravitational force between two protons (an enormous \$10^{36}\$) would catastrophically impact atomic stability. If slightly smaller, electrons would bind too tightly, preventing hydrogen formation. If larger, atomic nuclei might be unstable, or electrons might orbit too closely for stable quantum configurations.

The masses of fundamental particles are also critical. Up and down quark masses are crucial for proton and neutron stability. Altering their mass difference would jeopardize atomic nuclei stability, impacting nucleosynthesis and stable matter's existence. The weak nuclear force, vital for radioactive decay and stellar helium formation, also exhibits fine-tuning. Its strength dictates proton-to-neutron conversion rates in stellar cores, essential for stellar energy and heavier element production. A significantly different weak force might have caused the Sun to collapse or never ignite.

Perhaps the most striking example of fine-tuning is the cosmological constant, Λ, which governs the universe's accelerated expansion. Quantum field theory predicts an enormous vacuum energy density from quantum fluctuations. Calculations yield a staggeringly large vacuum energy, exceeding the observed dark energy value by an astonishing 10^{120}. This is the cosmological constant problem, arguably science's largest theory-observation discrepancy. If vacuum energy were near its predicted value, the universe would have expanded exponentially from the outset, preventing any structure formation. The observed value is so small, yet precisely non-zero, that it allows for gradual structure formation over billions of years while simultaneously driving current accelerated expansion.

The concept of fine-tuning, where numerous physical constants appear improbably calibrated for life, has often been addressed by the anthropic principle. This principle suggests we observe such a universe because, if it were otherwise, we wouldn't exist to observe it. While logically sound, it functions as a selection effect, describing a correlation without offering a mechanistic cause for this cosmic fine-tuning.

However, the Black Hole Big Bang hypothesis offers a more fundamental explanation. This theory proposes that our universe originated from within a black hole through a quantum gravitational transition. If this transition governs the fundamental constants and laws, then these parameters might not be arbitrary but established during this very process.

Under the extreme conditions within a black hole's event horizon, where our current physics breaks down and quantum gravity dominates, the universe's fundamental parameters could have been imprinted. This black hole acts as a "cosmic crucible," subjecting spacetime and energy to immense pressures and quantum fluctuations. The subsequent "bounce" into our expanding universe could then act as a natural selection mechanism, favoring specific values for these constants.

Quantum gravity theories, like loop quantum gravity and string theory, offer ways to conceptualize this, replacing the singularity with a region of finite curvature or smoothing it out. In these scenarios, the transition from a black hole to an expanding universe is a quantum event, a phase transition at the Planck scale. During this transition, fundamental symmetries could break, and physical constants could be determined.

If the Black Hole Big Bang model holds, physical laws and constants aren't arbitrary parameters assigned at a singular Big

Bang. Instead, they emerge from the specific quantum gravitational dynamics of the black hole-to-universe transition. Our universe would then represent the successful "settling" of these quantum gravitational degrees of freedom into a stable and complex state.

This transforms the fine-tuning problem from a matter of improbable luck into a question of the underlying physics of the black hole bounce. The specific values that emerge might be dictated by the stability of the resulting spacetime. Only certain constant combinations may allow a universe to expand and evolve without immediate collapse or dissipation. The observed fine-tuning would thus demonstrate the inherent stability of the spacetime configuration that emerged from the black hole.

The sheer number of finely tuned parameters further supports this argument, suggesting a deeper, unifying principle rather than coincidences. The Black Hole Big Bang scenario provides a single, complex event capable of establishing multiple parameters simultaneously. The dispersal of energy and matter from the black hole, viewed as a complex wave function collapse, with its outcome dictated by quantum gravity, would determine the nascent universe's fundamental properties.

The cosmological constant is a prime candidate for being determined by this process. If dark energy is a residual property of the black hole's spacetime geometry, its value would be intrinsically linked to the transition's physics. The significant discrepancy in vacuum energy calculations suggests an incomplete understanding or a dominant contribution from a source not captured by standard quantum field theory. The Black Hole Big Bang offers this source: the inherent spacetime tension or residual gravitational potential energy of the progenitor black hole. The quantum gravitational "bounce" could naturally "set" this tension to a non-zero but sufficiently small value for cosmic structure formation.

Furthermore, the emergence of dimensionality and spacetime properties could stem from this transition. Theories suggest extra dimensions at the Planck scale, and the Black Hole Big Bang could involve their compactification or shedding, leaving a 3+1 dimensional spacetime with specific properties. The resolution of these dimensions and the manifestation of forces could be intricately linked to the initial conditions and quantum dynamics of the bounce, influencing fundamental constant values.

Even the "cusp-core" problem in dark matter distribution, where simulations predict denser central regions than observed in some galaxies, might be related. If dark matter is a residual effect of the black hole's dispersal rather than an exotic particle, its initial distribution might have been more "cored" or its interactions with other bounce-generated components could have smoothed central densities. This implies dark matter's properties are not solely a consequence of its particle nature but also its cosmic origin from a non-uniform, anisotropic black hole dispersal.

The "fine-tuning problem" offers strong indirect support for a Black Hole Big Bang (BHBB) scenario. Instead of relying solely on luck or the anthropic principle, this hypothesis suggests a physical mechanism for the natural selection of the universe's fundamental parameters. The universe wouldn't be "designed" for life in a teleological sense; rather, its creation from a black hole would inherently favor a stable, evolving cosmic structure capable of supporting complexity. The constants would be fine-tuned out of necessity—the necessity of a stable, expanding spacetime emerging from the extreme conditions of a collapsed object. Their precise values would thus be a direct signature of the quantum gravitational physics at play during the universe's birth. This perspective shifts the focus from statistical improbability to the fundamental physics of cosmic genesis, offering a compelling, though speculative, resolution to a profound cosmological puzzle. Understanding these constants might therefore lead us directly to the secrets within a black hole's heart.

While the standard model of cosmology effectively describes the universe's evolution, it encounters observational puzzles regarding galaxy formation and distribution. The BHBB hypothesis could reinterpret these anomalies as signatures of an origin rooted in the dispersal of matter from an event horizon. These deviations, instead of being disparate unexplained phenomena, might be faint echoes of a singular, anisotropic origin event.

One such anomaly is the universe's large-scale structure—a cosmic web of filaments and voids. Although the Lambda-CDM (ΛCDM) model can generate such structures through gravitational instability acting on initial density fluctuations, their precise configuration and scale sometimes pose challenges. For instance, the observed distribution of galaxy clusters shows correlations extending to surprisingly large scales, questioning the homogeneity and isotropy assumptions of the standard model. If our universe originated from the highly asymmetric dispersal of matter from a black hole, the initial conditions might not have been perfectly isotropic. The outflow from an event horizon, influenced by the black hole's spin and mass, could naturally imbue the nascent universe with primordial anisotropy. This anisotropy, imprinted early on, might manifest as preferred directions or alignments in large-scale matter distribution, potentially explaining observed correlations that extend beyond predictions from purely random, isotropic fluctuations. Imagine the universe's initial "burst" not as an equal outward explosion, but as a directional expulsion, like a jet from an accretion disk, scaled cosmically. Such a process would inherently introduce preferred orientations into the primordial matter distribution, influencing subsequent structure growth.

The process of structure formation, where gravity amplifies small density fluctuations, might also carry imprints of this anisotropic origin. In the standard model, primordial fluctuations are assumed to be Gaussian and nearly scale-invariant, generated during cosmic inflation. However, if these fluctuations resulted from the quantum gravitational processes that birthed the universe from a black hole, their statistical properties might deviate from perfect Gaussianity. Furthermore, the matter distribution at the "bounce" moment might not have been perfectly uniform. A black hole's inherent structural asymmetry, even after a quantum transition, could lead to a non-uniform initial density field, affecting how galaxies and galaxy clusters form and evolve. For example, regions initially denser due to anisotropic dispersal might have seeded larger-than-expected superclusters or filaments, while regions with less initial material might have developed into cosmic voids. The specific clustering patterns, void sizes, and filament density profiles could all hold clues to the initial matter dispersal from the black hole.

The observed "cold spots" and "hot spots" in the Cosmic Microwave Background (CMB) radiation, while largely explained by standard cosmology, occasionally hint at potential anomalies. Some analyses suggest a statistically significant alignment of certain large-scale CMB features or a surprising lack of large-scale power. While typically attributed to statistical fluctuations within the standard model, the BHBB offers an alternative interpretation. The universe's initial state, emerging from the highly compressed and warped spacetime of a black hole, might not have been perfectly homogeneous in temperature or radiation. Residual anisotropies from the black hole's structure or specific quantum gravitational processes during the bounce could have imprinted a subtle, large-scale temperature gradient or non-Gaussianity onto the early universe's radiation. These imprints, stretched to cosmological scales by expansion, could manifest as apparent alignments or unusual temperature variations in the CMB. The progenitor black hole's spin, for instance, could directly influence the resulting quadrupole or octopole moment of the CMB, potentially explaining observed patterns statistically unlikely in the standard isotropic initial condition scenario.

The distribution and evolution of dark matter also present a relevant area. The standard ΛCDM model relies on dark matter as crucial scaffolding for galaxy formation. However, observations like the "cusp-core" problem (simulations predict denser dark matter cores in galactic centers than observed) and the "missing satellites" problem (simulations predict more small satellite galaxies around larger galaxies than observed) suggest our understanding of dark matter, its interaction with baryonic matter, and its initial distribution might be incomplete. If dark matter's origin is tied to the BHBB, its initial distribution could naturally differ from that produced by purely inflationary density fluctuations. For example, the energetic processes involved in matter dispersal from a black hole might have created a specific initial dark matter profile, potentially leading to less concentrated cores in nascent galaxies. Similarly, satellite galaxy formation might have been influenced by anisotropic outflow, leading to a different distribution than predicted by standard hierarchical clustering. Dark matter's very nature could be a relic of the universe's initial state, imprinted by its black hole progenitor's physics.

The observed properties of galaxies themselves might also offer subtle hints. While current models generally explain the typical sizes, shapes, and rotation curves of galaxies, outliers and statistical trends remain active research areas. For instance, extremely massive or compact galaxies, or unusual dynamical properties in some galactic systems, could potentially be linked to their formation history in a non-uniformly seeded universe. If the initial density field from which galaxies formed had inherent large-scale anisotropies or non-Gaussianities, it might have favored the formation of certain galaxy types in specific early universe regions, or influenced their subsequent evolutionary pathways. The presence of peculiar velocity flows, where galaxies move in coordinated patterns over vast distances, could also be more readily explained if the universe's initial conditions possessed preferred directions or large-scale velocity gradients, as might be expected from a directional cosmic event.

The universe's large-scale structure, known as the cosmic web, might be a direct result of a Black Hole Big Bang that produced a clumpy and filamentary initial matter distribution, rather than a smooth, homogeneous outflow. This non-uniformity could stem from the black hole's quantum state fragmenting or ejecting matter in distinct streams. This fundamentally different initial scaffold would then have shaped the subsequent gravitational collapse and evolution, with high-density outflows forming filaments and regions with less material becoming voids. The statistical properties and geometry of these initial ejections would thus be encoded in the observed distribution of galaxies and galaxy clusters.

Quantum gravity, though not fully understood, underpins the "bounce" from a black hole singularity. Theories like loop quantum cosmology and string theory models suggest a structured initial state for the universe, not merely random fluctuations. If our universe originated from a black hole, the quantum gravitational process would have imprinted unique characteristics onto the nascent cosmos. These characteristics, rather than being smoothed out, could persist as subtle deviations from predictions based on purely isotropic and homogeneous initial conditions.

This framework also offers a natural explanation for the observed coherence in galaxy spin alignment over large scales, a phenomenon difficult to explain with the standard model. If the initial outflow from the black hole possessed angular momentum or rotational shear, this "primordial spin bias" could have been inherited by emerging structures, leading to the observed alignments. The black hole's own spin and the anisotropic nature of the expulsion would naturally impart a preferred orientation to the aggregating matter.

The existence of ultra-diffuse galaxies (UDGs), which are as large as the Milky Way but contain few stars, is another observational puzzle. Standard models struggle to explain their low surface brightness and diffuse nature. However, non-uniform initial conditions from the black hole dispersal, with regions of lower density or different particle interactions, might have favored the formation of such diffuse structures. An extended initial distribution of dark matter or different initial dynamical influences on baryonic matter could have prevented denser galaxy formation. The specific energetic release or particle interactions during the Black Hole Big Bang could have created the conditions necessary for these peculiar, low-density galactic systems.

The study of large-scale structures also includes voids—vast, underdense regions surrounded by galaxy filaments and clusters. While the ΛCDM model predicts voids, their observed sizes and emptiness sometimes challenge simulations. A more structured initial state, with matter expelled in distinct streams, could naturally lead to larger and more pronounced voids. The geometry of these initial outflows would directly dictate the voids' size and shape, much like a jet from a spinning object creates high-density regions along its axis and underdensities perpendicular to it. The cosmic web would then be a direct consequence of this structured, anisotropic expulsion.

In summary, despite the standard cosmological model's success, anomalies in galaxy formation and distribution—including large-scale structure alignments, CMB patterns, dark matter profiles, and galaxy spin orientations—remain active research areas. The Black Hole Big Bang hypothesis offers a speculative, unifying framework for reinterpreting these puzzles. If the universe originated from a black hole's dispersal of matter, the inherent anisotropies and specific quantum gravitational processes could have imprinted patterns on the early universe. These patterns would influence galaxy and large-scale structure formation and distribution in ways that deviate from predictions of a purely isotropic and homogeneous origin. Observing these deviations, though subtle, could serve as indirect clues pointing towards a Black Hole Big Bang as the universe's ultimate genesis.

The search for these subtle imprints in the cosmic tapestry remains a frontier of astrophysical research, aiming to connect the cosmos' grandest structures with physics' most extreme environments.

Einstein's general relativity describes spacetime as being rippled by massive objects' motion, propagating as gravitational waves carrying information about their violent origins. While gravitational waves from black hole and neutron star mergers have been detected, the Black Hole Big Bang hypothesis suggests a grander, more primordial source. The universe's transition from within a black hole's event horizon to its current vast, expanding state would have been an event of unimaginable violence and energetic release. This cataclysmic interplay of quantum gravity and extreme spacetime curvature could, in principle, generate a unique spectrum of gravitational waves.

Unlike localized events like black hole mergers, a Black Hole Big Bang would have been a universe-scale event. The singularity within a black hole could be a transitional phase, not an endpoint. Quantum gravitational processes allowing the universe to "bounce" or emerge from this state would have involved the rapid conversion of immense gravitational energy into spacetime expansion and the creation of matter and radiation. This dramatic transformation would inevitably have set spacetime into oscillation, generating gravitational waves that permeated the nascent universe. The scale of this event would dwarf anything previously observed, a cosmic symphony of spacetime distortions from the very moment of creation.

The specific characteristics of these primordial gravitational waves would depend critically on the transition's nature from the black hole state to the expanding universe. A rapid phase transition or quantum fluctuation that "unwound" the singularity, for example, could have released energy in a burst, producing a characteristic gravitational wave signal with specific frequency content and amplitude. The progenitor black hole's nature—its mass, spin, and even its quantum state—could also leave an imprint. A rapidly spinning black hole might impart angular momentum or anisotropy to the outgoing gravitational wave signal, a subtle signature distinguishable from conventional astrophysical sources.

Furthermore, the very early universe, immediately after such a transition, would have been a highly energetic and dynamic environment. The quantum gravitational processes governing spacetime's birth might have produced a stochastic background of gravitational waves—a continuous hum of spacetime ripples rather than discrete events. This primordial gravitational wave background would be a direct fossil of the quantum gravity epoch, similar to how the Cosmic Microwave Background radiation is a fossil of a later era when the universe became transparent to photons. The frequency and amplitude of this background radiation would be exquisitely sensitive to the physics at play during the Black Hole Big Bang. If the transition involved quantum oscillations or fluctuations of the spacetime metric, these would manifest as a broadband gravitational wave signal, potentially extending across a wide range of frequencies.

Future gravitational wave observatories face the challenge of distinguishing primordial gravitational waves from the "cacophony" of astrophysical sources. However, the energy densities of the Big Bang are expected to be far higher than those from stellar-mass black hole mergers, suggesting primordial waves might occupy different frequency bands or possess unique statistical properties. For example, a Black Hole Big Bang could have generated gravitational waves in the nanohertz or microhertz range, currently inaccessible to ground-based detectors but targets for future space-based observatories like LISA. These low-frequency waves, stretched by cosmic expansion, would carry information from the earliest moments of existence.

The spectrum of gravitational waves from a Black Hole Big Bang might also exhibit distinct features. While standard inflationary cosmology predicts a specific spectrum with a characteristic "tilt" and amplitude, a Black Hole Big Bang scenario could predict a different spectrum. Depending on the transition mechanism, the resulting gravitational waves might be "blue-tilted" (more power at higher frequencies) or show specific oscillatory patterns not predicted by standard inflation. Such deviations would be crucial evidence for a Black Hole Big Bang origin.

Gravitational waves also offer a way to investigate dark matter and dark energy. If these phenomena are linked to the initial conditions set by a Black Hole Big Bang, their effects might be subtly encoded in the primordial gravitational wave signal. For instance, if dark matter originated from specific quantum excitations during the Black Hole Big Bang, these excitations might have produced a characteristic gravitational wave signature correlating with today's dark matter distribution. Similarly, if dark energy is a manifestation of the vacuum state emerging from the Black Hole Big Bang, the gravitational waves generated during that transition could provide clues about its nature.

Detecting these elusive primordial gravitational waves demands extraordinary sensitivity and sophisticated data analysis. Future ground-based and space-based observatories are designed to probe these early universe signals. LISA, with its long arm lengths, will be sensitive to millihertz to hertz frequencies, potentially capturing signals from phase transitions or other energetic events in the very early universe. Pulsar Timing Arrays (PTAs) are already probing the nanohertz regime and could detect the gravitational wave background from supermassive black hole mergers or even signals from a Black Hole Big Bang.

Comparing gravitational wave signals predicted by a Black Hole Big Bang scenario with those from standard inflationary cosmology will be crucial. If future observations reveal a gravitational wave background significantly deviating from inflationary predictions, especially in its spectral shape or polarization, it could provide strong evidence for alternative cosmological models, including the Black Hole Big Bang. For example, non-Gaussian features or a different power-law index from inflationary predictions might indicate a more complex or anisotropic origin. The polarization of gravitational waves also holds vital clues; deviations from predicted inflationary patterns could point to a different origin.

The potential for gravitational waves to probe the quantum gravity regime is perhaps their most exciting promise. The Black Hole Big Bang hypothesis places the universe's origin squarely within this enigmatic domain. If the universe emerged from a black hole, the gravitational waves produced would be direct messengers from this quantum realm, carrying information about spacetime's fundamental constituents, singularities, and quantum gravity laws. Detecting such waves would revolutionize cosmology and provide empirical support for quantum gravity theories, offering a direct window into the moment of creation.

The stochastic background of gravitational waves, if it exists, would be a continuous hum containing information about the universe's earliest moments. Unlike distinct signals from black hole mergers, this diffuse background would require long integration times and sophisticated statistical analysis to extract its properties. Future gravitational wave detectors, with increased sensitivity and wider frequency coverage, will be instrumental in mapping this primordial background with unprecedented precision, searching for characteristic signatures that might distinguish a Black Hole Big Bang from other cosmological paradigms. Listening to these oldest whispers of the universe through gravitational waves could fundamentally alter our perception of cosmic origins.

In essence, gravitational waves offer a unique and powerful probe into the universe's most extreme events and earliest moments. While current observations focus on astrophysical sources, the theoretical framework of a Black Hole Big Bang suggests that our universe's genesis could have been a gravitational wave-generating event of unparalleled magnitude. Future observatories, pushing technological boundaries, promise to detect these primordial ripples in spacetime. If such signals are found and bear the hallmarks predicted by the Black Hole Big Bang hypothesis—like a unique spectral shape, specific polarization patterns, or correlations with other cosmological phenomena—it would provide compelling, albeit indirect, evidence for a universe born not from a point of infinite density in empty space, but from the highly energetic and quantum-unstable heart of a black hole. This would represent a paradigm shift in our understanding of cosmic origins, connecting familiar aspects of our universe to the most exotic and enigmatic objects in the cosmos: black holes. The quest to detect these primordial gravitational waves is, therefore, a quest to understand not just the universe's past, but its very essence.

CHAPTER 7

Quantum Gravity and the Black Hole Boundary

Our current understanding of the cosmos rests upon two magnificent pillars: Einstein's General Relativity and Quantum Mechanics. General Relativity describes gravity not as a force, but as a curvature of spacetime, influencing the movement of planets and galaxies, and the nature of black holes. Quantum Mechanics, on the other hand, describes the universe at its most fundamental level, a realm of probabilities and discrete energy packets, governing the behavior of subatomic particles and the forces within atoms. Both theories have been extensively validated and form the bedrock of modern physics.

However, these two theories, while successful in their respective domains, clash when confronted with extreme environments like the singularity of a black hole or the initial moments of the Big Bang. General Relativity breaks down at these points, predicting nonsensical infinities. Similarly, at these scales, the smooth spacetime of General Relativity is expected to become a turbulent quantum foam, where Quantum Mechanics should prevail. Yet, attempts to apply Quantum Mechanics directly to gravity lead to profound mathematical difficulties and uncontrollable infinities. This represents a fundamental crisis in theoretical physics.

The Black Hole Big Bang hypothesis, which posits our universe emerged from within a black hole, directly addresses this conflict. To understand such a transition, we must reconcile gravity with quantum mechanics. We need to answer what happens when spacetime curvature becomes so extreme that quantum effects dominate, and how matter and energy behave when compressed into regions smaller than the Planck length. A complete theory of quantum gravity is essential to move this fascinating idea beyond speculation.

Physicists have grappled with the need for quantum gravity for decades, exploring various avenues for a unified description. Theories like String Theory and Loop Quantum Gravity offer radically different approaches. String Theory suggests fundamental particles are tiny vibrating strings, with gravity arising from the vibrations of a closed string. Loop Quantum Gravity proposes that spacetime itself is quantized, composed of discrete fundamental units or "loops." These and other frameworks aim to provide a consistent description of gravity at the Planck scale, where quantum gravitational effects are expected to be dominant.

The Black Hole Big Bang scenario offers a compelling reason to accelerate the development and testing of quantum gravity theories. If the universe did originate from a black hole, the physics governing that transition would be precisely that of quantum gravity. The characteristics of the emergent universe, including the spectrum of primordial gravitational waves, would bear the signature of this quantum gravitational process. For example, how a singularity is "resolved" or "avoided" in a quantum gravity theory would directly influence the Big Bang's nature. Whether the singularity quantum mechanically "bounces" or tunnels through a barrier, or if it expands uniformly or exhibits anisotropic behavior, the answers, hidden within quantum gravity equations, would manifest as observable features in gravitational wave data.

General Relativity describes the singularity within a black hole as a point of infinite density and curvature, offering no escape. However, quantum mechanics suggests that at such extreme densities, where particles are compressed to volumes smaller than their wavelengths, a classical description of spacetime must fail. Quantum fluctuations would become so violent that the very concept of smooth, continuous spacetime might break down. Quantum gravity theories attempt to describe this breakdown and the potential "resolution" of the singularity. Some theories propose that the singularity might be replaced by a region of extremely high, but finite, density and curvature, governed by quantum gravitational effects.

This "quantum core" could then be the seed from which a new universe emerges.

The nature of this quantum core is where quantum gravity theories diverge, and their predictions for a Black Hole Big Bang would differ. In some models, extreme compression could lead to a phase transition of spacetime itself—a violent rearrangement of its fundamental quantum structure. This transition could release a tremendous amount of energy, driving the expansion of the new universe and producing a characteristic gravitational wave signature. The specific details of this phase transition—its speed, energy scale, and underlying quantum degrees of freedom—would all be encoded in the gravitational waves. For instance, if the transition involved the creation of specific quantum excitations or the breaking of certain symmetries, these would leave a unique imprint on the gravitational wave spectrum, potentially distinguishable from signals predicted by standard inflationary cosmology.

Furthermore, the information paradox associated with black holes, which questions what happens to information that falls into a black hole, might find its resolution within a quantum gravity framework that also describes a Black Hole Big Bang. If a black hole can, in a sense, "reproduce" or "birth" a new universe, then the information that fell into the original black hole might be scrambled and re-emitted into the new cosmos. This process would inherently involve quantum gravitational effects, as the event horizon and singularity are central to the paradox. Understanding how information is preserved or transformed during this transition would require a complete quantum theory of gravity, and the signature of this information transfer could potentially be imprinted on the gravitational waves produced.

The very existence of the Planck scale, approximately 1.6 x 10^-35 meters and 1.22 x 10^-32 seconds, underscores the necessity of quantum gravity. At these infinitesimal scales, energy density and spacetime curvature are so immense that quantum effects are expected to dominate gravity's behavior. General Relativity, a classical theory, cannot describe physics at these scales. Quantum mechanics, while describing the quantum world, does not inherently include gravity. The Planck scale is the frontier where these two domains meet, and it is here that a theory of quantum gravity is essential for understanding phenomena such as the Big Bang or the interior of black holes. The Black Hole Big Bang hypothesis places the origin of our universe squarely within this Planckian regime.

The challenges in developing and experimentally verifying a theory of quantum gravity are immense. The energy scales at which quantum gravitational effects become significant are far beyond the reach of any current or foreseeable particle accelerator. This is why gravitational waves, particularly those from the very early universe, are so crucial. They offer a potential window into these inaccessible energy regimes. If the Black Hole Big Bang hypothesis is correct, then the gravitational waves generated during the transition from a black hole to our expanding universe would be direct probes of quantum gravity. The frequency, amplitude, and spectral shape of these primordial gravitational waves would be dictated by the specific quantum gravitational mechanisms at play.

For instance, if a quantum gravity theory predicts that singularities are smoothed out into a finite-sized quantum region, the size of this region and the dynamics within it would determine the gravitational wave output. A theory that predicts a "bounce" mechanism, where the collapsing matter within a black hole reaches a minimum size and then re-expands, would likely generate a specific pattern of gravitational waves. Conversely, a theory that allows for quantum tunneling through a barrier might produce a different gravitational wave signal. The ability to detect and analyze these signals would thus provide a direct empirical test for different quantum gravity frameworks, allowing us to move beyond purely theoretical speculation and gather observational evidence for the fundamental nature of gravity at its most extreme.

The pursuit of a quantum theory of gravity is inextricably linked to the search for a "theory of everything," a unified framework encompassing all fundamental forces: gravity, electromagnetism, and the strong and weak nuclear forces. If our universe emerged from a black hole, this event would represent a moment of profound unification, where the quantum gravitational structure of spacetime dictated the genesis of these forces and their mediating particles. A successful theory of quantum gravity would not only elucidate the Big Bang and the nature of black holes but also illuminate the unification of forces and the fundamental constituents of reality.

Thus, the Black Hole Big Bang hypothesis provides a compelling empirical motivation for tackling the formidable theoretical challenge of quantum gravity. It transforms the abstract quest for a unified theory into a concrete endeavor to understand our cosmic origins. The gravitational wave spectrum predicted by various quantum gravity models in the context of a Black Hole Big Bang could differentiate between them. For instance, if string theory forecasts a specific gravitational wave signal with unique polarization from the vibrational modes of strings during the transition, while loop quantum gravity predicts a different signal related to the discrete structure of spacetime, detecting such signals would provide strong evidence for one theory over the other.

Ultimately, quantum gravity is profoundly necessary for understanding the Black Hole Big Bang. It is the missing link connecting the macroscopic realm of gravity, described by General Relativity, with the microscopic world of quantum mechanics. Without it, our comprehension of how a universe could emerge from the extreme conditions within a black hole remains incomplete. Developing a consistent and predictive theory of quantum gravity is not merely an academic exercise; it is essential for answering fundamental questions about the nature of reality and our place within it. The universe's potential origin as a "quantum jump" from within a black hole makes this quest vital for understanding both extreme physics and our cosmic genesis.

The event horizon of a black hole, far from being the placid barrier suggested by classical General Relativity, is a region bustling with quantum activity. It is a boundary where the smooth, predictable fabric of spacetime is expected to unravel, revealing the universe's most fundamental and often counterintuitive laws. Exploring the quantum realm surrounding this enigmatic threshold uncovers phenomena that challenge our everyday understanding of reality and hold significant implications for the Black Hole Big Bang hypothesis.

One of the most renowned quantum effects associated with black holes is Hawking radiation, the theoretical emission of particles from the event horizon. Elucidated by Stephen Hawking, this phenomenon stems from the ceaseless quantum fluctuations that pervade the vacuum of spacetime. According to quantum field theory, even empty space is a dynamic arena where particle-antiparticle pairs spontaneously appear and then annihilate moments later. These virtual particle pairs, a manifestation of the Heisenberg uncertainty principle, are constantly created and destroyed throughout the universe.

However, at the event horizon, these virtual particles face a unique and critical situation. When a virtual particle-antiparticle pair emerges precisely at the boundary, one particle might fall into the black hole while the other escapes. The particle falling in is inevitably directed towards the singularity, lost to the outside universe. Conversely, the escaping particle is directed outwards, carrying energy away from the black hole. To a distant observer, it appears as though the black hole itself is radiating particles and, consequently, energy. This emitted radiation is thermal, possessing a blackbody spectrum, and its temperature is inversely proportional to the black hole's mass. Smaller black holes are hotter and radiate more intensely, while larger ones are colder and radiate more feebly.

The origin of this emitted energy is crucial for understanding its role in the Black Hole Big Bang scenario. It is not created ex nihilo; rather, it is drawn from the black hole's gravitational energy. A virtual particle falling into the black hole can be thought of as carrying negative energy, thereby reducing the black hole's total mass and energy. The escaping particle, conversely, carries away positive energy, forming Hawking radiation. Thus, Hawking radiation leads to the slow but steady evaporation of black holes over vast timescales.

The implications of Hawking radiation for the Black Hole Big Bang hypothesis are diverse. If our universe originated within a black hole, the very process of this transition would have been deeply intertwined with quantum phenomena at the event horizon. The energy that powered the Big Bang—the expansion of spacetime and the creation of all matter and energy in our universe—might have been intimately connected with the quantum processes occurring at the boundary of the parent black hole. It is conceivable that the energy released through Hawking radiation, or a more generalized quantum process occurring at the event horizon during a critical phase, provided the initial impetus for cosmic expansion.

Consider the extreme conditions near the event horizon. As matter and energy collapse to form a black hole, they are compressed to densities and curvatures that General Relativity struggles to describe. While the event horizon itself is a classical concept, the quantum vacuum at this boundary is far from quiescent. The intense gravitational field warps spacetime so dramatically that the conventional notion of particles and fields begins to break down. Virtual particle pairs are not merely fleeting excitations; their creation and separation become a fundamental aspect of the physics at the horizon. The energy associated with these quantum fluctuations can manifest as real particles that escape to the outside.

In the context of a black hole giving birth to a universe, the event horizon would not merely be a point of no return for infalling matter but a dynamic interface where the information and energy contained within the black hole are processed and potentially transformed. The quantum fluctuations near the horizon could have played a crucial role in this transformation. Instead of simply radiating away as thermal particles, the quantum vacuum energy might have been channeled or amplified in a manner that led to the explosive expansion characteristic of the Big Bang.

This brings us to the perplexing concept of the information paradox. According to quantum mechanics, information cannot be destroyed. However, Hawking radiation, being thermal, appears to be purely random and devoid of any information about the matter that originally formed the black hole. If a black hole eventually evaporates completely into Hawking radiation, all information about its interior would seemingly be lost, violating a fundamental tenet of quantum theory.

The Black Hole Big Bang hypothesis offers a potential, albeit speculative, resolution to this paradox. If the black hole does not fully evaporate but instead transitions into a new universe, then the "lost" information might not be destroyed but rather scrambled and re-emitted into the nascent cosmos. The event horizon, as the site of this transformation, would be the critical region where the quantum processes responsible for both Hawking radiation and the potential universe-birth occur. Understanding how information is preserved or encoded within this process is a key challenge for any quantum theory of gravity.

The exact process by which quantum fluctuations at a black hole's event horizon could initiate a Big Bang is a subject of ongoing theoretical research. One theory suggests that in the final stages of a black hole's existence, as it shrinks and its Hawking temperature intensifies, quantum effects become so dominant that they fundamentally transform the event horizon. This smooth boundary might instead become a turbulent, highly energetic quantum region. It is within this "quantum cauldron" that the conditions necessary for a new universe to emerge could be met, with immense energy density potentially triggering a phase transition of spacetime itself.

Envision the event horizon not as a static barrier, but as a dynamic quantum frontier. When the matter forming the black hole reaches its most compressed state, quantum vacuum fluctuations at this boundary could interact to produce a "quantum bounce" or a rapid phase transition. This transition could unleash a vast amount of energy, creating the conditions for a universe to inflate and expand. The resulting universe's radiation spectrum and inherent nature would bear the quantum imprints of this process.

Furthermore, if the Black Hole Big Bang hypothesis is correct, the initial quantum state of the new universe would be directly linked to the quantum state of the parent black hole's event horizon at the moment of transition. While seemingly random on average, the properties of Hawking radiation are believed to be governed by the quantum correlations of the vacuum. If these correlations are preserved and amplified during the transition, they could determine the initial conditions of our own universe, including the amplitude and spectrum of primordial density fluctuations, and possibly even the nature of the initial quantum gravitational field.

The quantum nature of the event horizon also raises questions about causality and determinism. In classical General Relativity, the event horizon is a causal boundary where events inside can influence those outside, but not vice versa. However, quantum mechanics introduces probabilities and uncertainties. The emission of Hawking radiation suggests a breakdown of classical determinism at the horizon. For a Black Hole Big Bang, this implies that our universe's origin might be rooted in fundamentally probabilistic quantum events occurring at the boundary of its parent black hole.

At the Planck scale—the incredibly small scale where quantum gravity effects are expected to dominate—spacetime itself is thought to be a "quantum foam." Near a black hole's event horizon, spacetime is already highly curved, and it's plausible that at its very edge, this quantum foam becomes the dominant feature. This foam would consist of constantly appearing and disappearing fluctuating regions of spacetime, within which the seeds of a new universe might have been sown. The energy required to manifest these fluctuations as real particles, or as the initial expansion of a universe, would ultimately be drawn from the black hole's mass-energy.

Theoretical models that describe matter and energy at extreme densities within a black hole, especially at its boundary, often draw upon concepts from quantum field theory in curved spacetime and ultimately highlight the need for a complete theory of quantum gravity. These theories suggest that the singularity predicted by General Relativity might be "smoothed out" by quantum effects, replaced by a region of extremely high, but finite, energy density and curvature. In this view, the event horizon becomes the gateway or interface to this quantum realm, and from this quantum core, a new, expanding spacetime could emerge.

The temperature of Hawking radiation is a direct result of gravitational redshift near the event horizon. As virtual particles separate, one falls into the black hole while the other escapes, gaining energy as it climbs out of the intense gravitational potential well. This energy acquisition is directly linked to quantum vacuum fluctuations in the presence of strong gravity. For the Black Hole Big Bang hypothesis, this magnified and transformed energy gain mechanism could be the source of the immense energy required for cosmic inflation.

Furthermore, the emission of Hawking radiation can be seen as a continuous "leakage" of quantum information and energy from the black hole. If this leakage, instead of simply dissipating as thermal radiation, were to channel into a process of spacetime creation, it would provide a mechanism for the Black Hole Big Bang. The rate of this leakage, determined by the black hole's mass and quantum gravitational properties, would then set the initial conditions for the expansion of the new universe.

The formation of a black hole and the subsequent emergence of a new universe from within are intricately tied to the quantum nature of spacetime itself. The event horizon, where these two regimes meet, is a hub of intense quantum activity. The quantum vacuum fluctuations that give rise to Hawking radiation are not merely an academic curiosity; they are a profound manifestation of the energetic and dynamic nature of spacetime at its most extreme. It is within this quantum realm, at the very edge of the abyss, that the potential for cosmic rebirth may lie, bridging the enigmatic nature of black holes and the grand origins of our own universe. The information paradox, unresolved within classical physics, finds its potential solution in the quantum processes at the event horizon, suggesting that information might be encoded in the very fabric of spacetime during this transition.

The classical depiction of a black hole, as dictated by Einstein's General Relativity, culminates in a terrifying abyss: the singularity. At the heart of every black hole, theory predicts a point of infinite density and zero volume, where the laws of physics as we know them cease to apply. Here, spacetime curvature becomes infinitely large, and all matter and energy that have crossed the event horizon are crushed into an infinitely small point. This theoretical construct, while a powerful consequence of the equations, presents a profound challenge to our understanding of the universe, particularly when considering its potential role in cosmic origins, as suggested by the Black Hole Big Bang hypothesis. If a black hole is the crucible from which a new universe emerges, then the nature of this singularity becomes paramount. However, the presence of such an infinite point also signals a breakdown in our current physical theories, indicating that General Relativity, which describes gravity and spacetime on large scales, is incomplete. To truly understand what happens at the heart of a black hole, and thus what might be the genesis of a universe, we must turn to a theory that can reconcile gravity with the quantum world: quantum gravity.

The singularity problem is not exclusive to black holes; it also appears in the Big Bang model of cosmology. The Big Bang theory, derived from General Relativity, suggests that the universe began from an initial state of infinite density and temperature. This cosmological singularity shares many conceptual similarities with the singularity at the center of a black hole. Both represent points where the gravitational field becomes overwhelmingly strong, leading to infinite spacetime curvature and the breakdown of classical physics. This shared feature is a significant reason why the Black Hole Big Bang hypothesis is so compelling. If a quantum theory of gravity can resolve the singularity in one context, it might offer a unified solution for both black holes and the origin of our universe. The key idea is that at these extreme densities and curvatures, quantum mechanical effects, which are negligible in everyday life, become dominant and fundamentally alter the nature of spacetime.

The pursuit of a consistent theory of quantum gravity is a central challenge in modern physics, aiming to unite gravity and quantum mechanics. This endeavor seeks to describe gravity at a quantum level and resolve singularities predicted by classical theories. Two prominent theoretical frameworks are String Theory and Loop Quantum Gravity (LQG), which, despite their differing approaches, share the goal of providing a more complete understanding of reality at its most fundamental level.

String Theory posits that the universe's fundamental components are not point-like particles but tiny, vibrating strings. These strings can be open or closed, with their various vibrational modes corresponding to different elementary particles, including the graviton. In String Theory, singularities are resolved by the finite size of these fundamental strings. Instead of collapsing infinitely, strings resist compression, vibrating energetically or expanding into new dimensions. This suggests that at extreme energies and densities, spacetime might exhibit a "fuzzball" or other complex quantum structure, replacing the singular point with a region of maximal, but finite, quantum excitation.

The implications of String Theory for a Black Hole Big Bang scenario are profound. If a black hole singularity is a region of finite quantum density rather than an infinitely dense point, it could act as a "bounce" point. Matter and energy would reach a maximum density and then rebound, initiating a new phase of expansion. This "quantum bounce" would be the equivalent of a Big Bang originating within the black hole, with energy and information preserved and reconstituted in the emerging universe. The behavior of D-branes or other extended objects within string theory, as well as the finite size of strings, could influence this bounce and potentially resolve the information paradox by allowing information to escape or be imprinted on Hawking radiation.

Loop Quantum Gravity (LQG) offers an alternative approach, proposing that spacetime is granular, composed of discrete, quantized units of volume and area, visualized as loops or networks. As matter collapses towards a black hole's center in LQG, it encounters these discrete spacetime quanta. Compression is halted at a minimum size determined by the Planck length and area, leading to a "quantum bounce."

In LQG, the singularity is replaced by a region of high, but finite, curvature and energy density, such as a "Planck-scale firewall." This quantum bounce mechanism supports the Black Hole Big Bang hypothesis, suggesting that a black hole transitions into a new, expanding universe rather than ending in a singularity. Collapsed matter and energy are compressed to a critical density, triggering a rebound that initiates a new cosmic expansion. The properties of this emergent universe would be shaped by the quantum gravitational processes during the bounce. The discrete nature of spacetime in LQG could also preserve information, as spacetime's quantum states would retain an imprint of what entered the black hole.

The concept of a "quantum bounce" resolving singularities has also been explored in Loop Quantum Cosmology (LQC), which applies LQG principles to cosmological models. LQC suggests that the Big Bang singularity can be replaced by a bounce, where a contracting universe transitions into an expanding one. Applied to black holes, this implies that the interior of a black hole is a region transitioning from collapse to expansion, potentially seeding a new universe when density reaches a maximum finite value.

The Black Hole Big Bang hypothesis, viewed through these quantum gravity theories, presents a compelling narrative for cosmic origins: our universe could be the offspring of a black hole in a parent universe. The "singularity" within this parent black hole would be a locus of intense quantum gravity effects. As matter and energy collapsed, they reached a critical density where quantum gravity intervened, preventing a true singularity and instead inducing a "bounce." This powerful bounce would have created our expanding universe. The parent black hole's event horizon, while appearing to trap everything from the outside, would have been a gateway to a new cosmic epoch from the inside. Information seemingly lost to Hawking radiation could, in this scenario, be encoded within the quantum gravitational state leading to the bounce and the subsequent creation of our universe.

From Black Hole Singularity to Cosmic Genesis

The resolution of a black hole's quantum singularity offers a compelling explanation for the universe's initial conditions. The specific characteristics of the quantum bounce—its density, curvature, and the quantum states of spacetime—would profoundly influence the emergent universe. For instance, the spacetime fluctuations during this bounce could be the origin of the primordial density fluctuations observed in the cosmic microwave background radiation, which seeded all large-scale structures in our universe. Even the fundamental properties of our spacetime, such as its dimensionality and constants, might have been determined by the quantum gravitational state of the parent black hole's core at its moment of rebirth.

The transition from a collapsing black hole interior to an expanding universe is an intricate process requiring a complete understanding of quantum gravity. Theories like String Theory and Loop Quantum Gravity offer promising pathways to resolve the singularity and facilitate such a transition, though the precise mechanisms are still under active investigation. These theories consistently suggest that the infinities predicted by General Relativity are artifacts of an incomplete theory. Instead, a quantum description of gravity would reveal a fundamentally smooth universe, albeit with discrete quantum structures at the Planck scale. The singularity, rather than an infinitely dense point, transforms into a region of extreme, finite quantum activity—potentially the very engine for cosmic creation. This framework re-envisions the black hole not merely as an object of cosmic destruction, but as a potential incubator for new beginnings, establishing a profound link between these enigmatic celestial bodies and the grand narrative of our existence.

The Black Hole Information Loss Paradox

The information loss paradox is a profound puzzle at the intersection of general relativity and quantum mechanics, specifically concerning black holes. It questions whether quantum information, representing the precise state of a system, is irrevocably destroyed upon entering a black hole or if it is somehow preserved and eventually re-emerges.

In classical physics, a black hole is an ultimate trap. Anything crossing its event horizon is drawn towards the central singularity. General Relativity starkly dictates that once matter, energy, and their associated quantum information fall into a black hole, they are lost to the outside universe. Classically, a black hole is characterized only by its mass, electric charge, and angular momentum—the "no-hair theorem." All other unique details of the infalling objects are seemingly erased.

The situation becomes far more complex with quantum mechanics and Stephen Hawking's discovery of Hawking radiation. Hawking demonstrated that black holes are not truly black; they emit thermal radiation due to quantum effects near the event horizon. This radiation is a stream of particles escaping the black hole's gravity. Crucially, Hawking's initial calculations suggested this radiation is purely thermal, carrying no information about the black hole's specific contents, much like how the photons from a uniformly heated object reveal its temperature but not its composition.

If a black hole completely evaporates via this information-free Hawking radiation, the information that originally fell in would be permanently lost. This directly contradicts unitarity, a fundamental principle of quantum mechanics. Unitarity implies that the evolution of a quantum system is reversible; in principle, its history can be traced from its final state. Information is conserved, not destroyed. The information loss in black hole evaporation thus represents a profound conflict between our two most successful physics theories.

The implications of this paradox are far-reaching. Information destruction would challenge the fundamental determinism described by quantum mechanics and suggest an incomplete understanding of causality and the flow of time, particularly in extreme gravitational environments. The very fabric of spacetime, as described by general relativity and modified by quantum effects, appears to be the battleground for this fundamental principle.

Numerous theoretical proposals aim to resolve the information loss paradox by reconciling gravity and quantum mechanics. One prominent class suggests information is subtly encoded within the Hawking radiation. While the radiation appears thermal on average, proponents argue for tiny, non-thermal correlations carrying the imprint of infalling matter. These correlations are extremely difficult to detect amidst thermal noise, but their existence would preserve unitarity. A key idea here is that Hawking radiation itself might be a quantum mechanical process involving entanglement between particles inside and outside the black hole. As a black hole evaporates, this entanglement could evolve, allowing information to leak out. Early proposals envisioned information being "scrambled" by the black hole's intense gravity and then re-emitted, akin to breaking down a complex recipe and then slowly reassembling its individual ingredients.

Another significant avenue involves the concept of complementarity, developed by physicists like Gerard 't Hooft and Leonard Susskind. This idea proposes that from the perspective of an infalling observer, information passes through the event horizon towards the singularity, albeit in a highly distorted form. However, from an external observer's perspective, the same information is imprinted on the event horizon and subsequently radiated away as Hawking radiation. The paradox arises from trying to reconcile these two seemingly contradictory but equally valid viewpoints. Complementarity suggests both descriptions are correct, but they cannot be simultaneously observed by the same person. Information isn't truly lost; it's simply accessed or described differently depending on the observer's frame of reference.

Quantum gravity theories propose radical solutions to the black hole information loss paradox. String Theory suggests black holes are "fuzzballs"—complex, extended quantum states without a singularity or sharp event horizon. Information would interact with this fuzzy boundary and be re-emitted, preserving details. The finite size of string constituents prevents infinite compression, thus avoiding an information-destroying singularity. The entanglement entropy of these fuzzballs could explain Hawking radiation while maintaining unitarity.

Loop Quantum Gravity (LQG) offers another perspective, viewing spacetime as granular. Inside a black hole, the singularity might be a Planck-scale region undergoing a quantum bounce. Information would pass through this bounce, potentially emerging in a new universe or being re-emitted via a quantum-structured horizon. This suggests quantum geometry, not classical continuum, dictates behavior near the classical singularity.

The "firewall paradox," proposed by AMPS, highlighted the tension between quantum mechanics and general relativity. It posited that if Hawking radiation carries information and is entangled with the black hole's interior, an infalling observer would encounter a high-energy "firewall" at the event horizon, contradicting general relativity's equivalence principle that freefall should be smooth. Solutions include "quantum entanglement teleportation" and more complex, non-local entanglement structures for Hawking radiation, aiming to smooth the horizon while allowing information escape.

The Black Hole Big Bang hypothesis provides a unique context. If our universe originated from a prior black hole's collapse, then information forming our universe must have traversed that parent black hole's event horizon. Information conservation, crucial for quantum mechanics, implies the parent black hole preserved and re-emitted the information seeding our cosmos. This suggests the quantum gravitational processes within a black hole leading to a new universe must be unitary. The very existence of our ordered, information-rich universe could be indirect evidence for the information loss paradox's resolution in favor of unitarity, with initial quantum states meticulously preserved through a quantum bounce or similar process.

Resolving the information loss paradox is vital for a complete quantum gravity theory. It serves as a crucial test for theoretical frameworks, demanding a consistent answer on information conservation in black holes. Success would deepen our understanding of black holes, spacetime, quantum mechanics, and the cosmos' fate. It also profoundly connects to the Black Hole Big Bang hypothesis, suggesting information's ultimate fate within a black hole could be rebirth into a new cosmic era, linking deep celestial mysteries to our universe's origin. The paradox's resolution holds the key to understanding both the fate of matter falling into black holes and the possibility of cosmic genesis through such mechanisms.

The conundrum of black hole information loss has spurred sophisticated theoretical frameworks to reconcile general relativity with quantum mechanics. Black hole complementarity and the holographic principle are potent ideas offering solutions and new perspectives on spacetime and information. These represent paradigm shifts, prompting a re-evaluation of fundamental assumptions about reality in extreme black hole environments.

Black hole complementarity, proposed by physicists like Gerard ‘t Hooft and Leonard Susskind, resolves the paradox of information loss in black holes by suggesting that the seemingly contradictory descriptions of information falling into a black hole are both valid, but from different perspectives.

Two Perspectives on Information:

- **Infalling Observer:** An observer falling into a black hole experiences nothing unusual as they cross the event horizon. Their information, consistent with the principle of equivalence, proceeds towards the singularity.
- **External Observer:** An observer at a safe distance sees information being encoded onto the event horizon and then gradually radiated away as Hawking radiation.

These two perspectives are complementary and cannot be simultaneously occupied by a single observer. Information is not lost; its description is simply observer-dependent. The paradox only arises when attempting to reconcile both viewpoints from a single observer, which complementarity argues is impossible due to the nature of quantum measurement and entanglement.

To illustrate, consider a particle falling into a black hole:

- **Infalling Observer:** The particle continues its trajectory into the interior.
- **External Observer:** The particle's quantum state becomes entangled with the emitted Hawking radiation. As the black hole evaporates, this entanglement builds, and the particle's information is subtly imprinted onto the outgoing radiation. The information is, in essence, "copied" or "distributed."

Complementarity asserts that no conflict exists because an observer cannot simultaneously be both inside and outside the black hole. Information is preserved, but its manifestation depends on the observer. This upholds unitarity (quantum information is never truly lost) without violating the principle of equivalence for the infalling observer. However, the precise mechanism of this "copying" and "distribution," especially regarding entanglement, remains an active research area.

The Holographic Principle: A Deeper Understanding

This concept of observer-dependent information leads to the holographic principle, a radical idea with profound implications for black hole interiors and the origins of universes. Inspired by black hole thermodynamics and string theory, it suggests that the description of a volume of spacetime can be encoded on its boundary, much like a 3D image on a 2D screen. For black holes, this implies all information, including the singularity, is fully described by degrees of freedom on the event horizon. The event horizon, classically a surface without thickness, becomes the locus of all information about the black hole's interior. This remarkable notion suggests that the complexity of a gravitational system can be reduced to a simpler, lower-dimensional system on its boundary.

Origins in Black Hole Entropy:

The holographic principle stems from the discovery by Jacob Bekenstein and Stephen Hawking that black hole entropy (a measure of disorder or information) is proportional to the *area* of its event horizon, not its volume. This was a significant departure from conventional physics, where entropy scales with volume. The area-dependent scaling suggests that information is stored on the black hole's surface, leading to the conjecture that information in *any* spacetime region might be encoded on its boundary. The Planck area (approx. 10^-70 square meters) represents the ultimate limit for this encoding, implying that spacetime is granular at the smallest scales and information density is finite.

Holography and the Information Loss Paradox:

Applied to black holes, the holographic principle offers a solution to the information loss paradox. If all information about fallen matter is encoded on the event horizon, then as the black hole evaporates via Hawking radiation, this information is gradually released. The Hawking radiation, originating from quantum effects at the horizon, would then carry this encoded information. The challenge lies in deciphering how this information is encoded and precisely transferred from the horizon to the emitted radiation. It suggests information isn't lost in a singularity but is stored on the boundary and "painted" onto the outgoing radiation.

String Theory's Contribution:

Advances in string theory have dramatically illustrated this idea. In certain string theory models, specific black holes can be described as collections of strings and D-branes. The statistical counting of these stringy states precisely reproduces the Bekenstein-Hawking entropy formula, providing strong evidence for the holographic principle. These black holes appear holographic, with their properties fully described by the dynamics of their constituent strings and branes. String-theoretic descriptions offer a microscopic origin for black hole entropy and suggest a mechanism for information preservation during black hole formation and evaporation.

Broader Implications:

The holographic principle extends beyond black holes, suggesting our entire universe might be holographic. This astonishing proposition implies that the fundamental degrees of freedom constituting reality could reside on a lower-dimensional boundary at the edge of the observable universe, or on a more fundamental spacetime manifold. Thus, the vast expanse of our cosmos, with all its galaxies, stars, and matter, could be a projection—a cosmic hologram.

The Black Hole Big Bang hypothesis, when viewed through the lens of the holographic principle and black hole complementarity, offers a compelling framework for understanding the origins of our universe and resolving the information loss paradox. This perspective suggests a profound connection: if our universe emerged from a black hole in a precursor universe, then the information that shaped our cosmos was holographically encoded on that parent black hole's event horizon.

In this scenario, as the parent black hole evaporated or underwent a quantum gravitational transition, it would have released this information, thereby initiating the Big Bang of our universe. The "interior" of the parent black hole, where our universe was conceived, is thus holographic – its entire content is fully described by its boundary, the event horizon. This means that information is not lost but preserved on this boundary and subsequently transferred to seed a new cosmic epoch.

This framework elegantly addresses the information loss paradox in the context of cosmic origins. Information falling into a black hole isn't destroyed; rather, it's transferred to the event horizon and can be released as Hawking radiation. If a black hole's ultimate destiny is to birth a new universe, then the information encoded on its horizon directly dictates the initial conditions of that nascent universe. From this viewpoint, the Big Bang isn't a creation *ex nihilo* but a re-emergence of preserved information – a cosmic rebirth facilitated by the quantum gravitational processes governing the black hole's boundary.

The concept of information scrambling within a black hole further strengthens this understanding, being deeply intertwined with both complementarity and holography. Black holes are considered the universe's most efficient scramblers of information. When matter enters a black hole, its quantum information rapidly mixes and entangles with the black hole's internal degrees of freedom, making its retrieval in its original form incredibly difficult. However, scrambling doesn't equate to erasure. Complementarity posits that this scrambling allows for the eventual release of information via Hawking radiation, while the holographic principle suggests that scrambling might be an emergent property of the degrees of freedom residing on the event horizon. The complexity of the black hole's interior is mapped onto its simpler boundary, and the dynamics on this boundary govern how information is processed and emitted.

Theoretical tools like the Anti-de Sitter/Conformal Field Theory (AdS/CFT) correspondence provide concrete mathematical evidence for the holographic principle. This powerful duality in string theory demonstrates an equivalence between a gravitational theory in Anti-de Sitter space and a quantum field theory without gravity on its boundary. This allows physicists to analyze strongly coupled quantum systems, which are typically challenging, by mapping them to more tractable gravitational systems in higher dimensions. The AdS/CFT correspondence serves as a working model where holography is a precise mathematical fact, significantly contributing to our understanding of black hole physics and the information paradox. By studying the CFT on the boundary, physicists can gain insights into the gravitational dynamics of the corresponding AdS black hole, including how information is processed and preserved.

Furthermore, the holographic principle offers a fresh perspective on the nature of singularities. Classically, the singularity at a black hole's center is a point of infinite density and curvature where the laws of physics break down. However, if all information about the black hole's interior is encoded on the event horizon, the singularity itself might be an emergent phenomenon or an artifact of our incomplete understanding of quantum gravity. In a holographic description, the "singularity" might not be a point of infinite density but rather a region where the holographic description becomes particularly complex, or where the boundary degrees of freedom interact in intricate ways. Some theories propose that the singularity is "resolved" by the holographic mapping, replaced by a smooth, non-singular description within the boundary theory.

If our universe is indeed holographic and originated from a black hole, it implies an intrinsically quantum mechanical and information-preserving universe from its very inception. The Big Bang would then be a phase transition – a manifestation of information emerging from a lower-dimensional substrate. The initial state of our universe and the finely tuned parameters governing its evolution would all be encoded in the properties of the precursor black hole's event horizon. This establishes a profound link between the quantum nature of spacetime, the enigma of black holes, and the origin of everything we observe. The idea that our universe could be a "projection" from a lower-dimensional reality, possibly born from the ultimate fate of a black hole, is one of the most mind-bending consequences of these advanced concepts. It suggests that the fundamental building blocks of our reality might not be particles and fields in three spatial dimensions, but rather the quantum information encoded on some fundamental boundary.

Ongoing research in this field is continually pushing the boundaries of our comprehension. While these ideas are highly theoretical and often inhabit the frontiers of mathematical and physical speculation, they offer a coherent and elegant approach to addressing some of the most profound puzzles in modern physics. The interplay between black hole complementarity and the holographic principle provides a powerful conceptual toolkit for exploring the deep connections between gravity, quantum mechanics, and information. These concepts suggest a universe where information is a fundamental currency, conserved and transformed through seemingly paradoxical processes, ultimately leading to the emergent reality we experience, and potentially, to the birth of new universes from the ashes of old black holes. The quest for a complete theory of quantum gravity is intrinsically linked to unraveling these mysteries, and these concepts serve as vital signposts on that journey. They are not merely abstract philosophical musings; they represent the logical extensions of our most successful physical theories pushed to their most extreme limits.

CHAPTER 8

Mathematical Models and Simulations

The theory that our universe originated from a black hole in a preceding cosmic epoch demands the rigorous development of mathematical models to evolve from a conceptual hypothesis into a robust scientific framework. This endeavor merges theoretical physics' abstract elegance with the stringent requirements of quantitative description. The core challenge lies in reconciling the smooth, deterministic world of black holes, as depicted by Einstein's general relativity, with the probabilistic, quantized realm of quantum mechanics. This is particularly difficult when considering the extreme conditions during a black hole's formation, evaporation, and the subsequent birth of a new universe.

Central to this effort is the need to quantify the proposed cosmic transformation. If a universe arises from an evaporating black hole, the process must be governed by equations that describe the interplay between gravity and quantum information as the black hole diminishes. General relativity, despite its success in describing large-scale spacetime and massive objects, fails at the black hole's singularity and the universe's earliest moments, where quantum gravitational effects are expected to dominate. Thus, the mathematical framework must integrate these quantum aspects while remaining consistent with general relativity in certain regimes.

A primary obstacle is developing a quantum description of spacetime, especially within quantum gravity. Theories like string theory and loop quantum gravity offer potential solutions. String theory posits that the universe's fundamental constituents are not point-like particles but tiny, vibrating strings. In some string theory models, black holes are understood as collections of these strings and higher-dimensional objects called branes. The precise mathematical description of these entities' interactions and dynamics could reveal how information is stored and released during a black hole's life cycle, potentially leading to a new universe. String theory's mathematical tools, including complex geometries, supersymmetry, and a vast landscape of solutions, provide fertile ground for constructing such models. For example, the AdS/CFT correspondence, a powerful duality in string theory, suggests that a gravitational theory in Anti-de Sitter space is equivalent to a quantum field theory on its boundary. This duality has been crucial for understanding black hole thermodynamics and demonstrating holographic systems, where information is encoded on boundaries. If our universe emerged from a black hole, the mathematical description of its evaporation and subsequent "bounce" into a new universe might be found within such holographic dualities, where complex internal dynamics are mapped to simpler boundary phenomena.

Conversely, loop quantum gravity quantizes spacetime itself, proposing a granular, atomic structure at the Planck scale, described by networks of interconnected loops. When applied to a black hole's interior, this framework suggests that the singularity is replaced by a region of high but finite curvature where spacetime "bounces" instead of collapsing infinitely. This quantum bounce mechanism is particularly appealing for the black hole-to-universe scenario: a collapsing matter cloud forms a black hole, which then undergoes a quantum bounce, expanding outwards to become a new universe. The mathematical challenge here is to accurately describe the quantum states of spacetime during this bounce, including the dynamics of matter and energy that would eventually form the new cosmos. This necessitates solving complex quantum differential equations that govern the evolution of these quantum gravitational states.

Mathematical models in physics must confront the information paradox, which suggests that if quantum mechanics is correct, Hawking radiation from evaporating black holes must carry information. For a universe born from a black hole, this implies that the new universe's initial conditions, fundamental laws, and constants would be encoded in the parent black hole's Hawking radiation. The challenge lies in mathematically demonstrating this encoding, potentially by studying quantum entanglement between the Hawking radiation and the nascent universe during evaporation. Quantum information theory techniques, such as entropy and mutual information calculations, are essential. For example, a model might aim to show how the quantum state of emitted radiation, precisely analyzed, contains all information about matter that fell into the black hole, including the seeds for new cosmic structures, through subtle correlations governed by complex quantum field theory calculations in curved spacetime.

Furthermore, any mathematical framework must account for our universe's observed properties. If our universe originated from a black hole, the model must reproduce the characteristics of the Big Bang, the cosmic microwave background radiation, the large-scale structure of galaxies, and the fundamental constants of nature. This necessitates integrating cosmology and particle physics into a quantum gravity framework. For instance, the early universe's inflationary epoch could be explained as a consequence of a quantum gravitational transition or specific properties of the boundary from which the universe emerged. Mathematical models would need to derive the equations of motion for fields driving inflation, demonstrating their origin from the dynamics of the parent black hole's horizon or interior.

Developing these mathematical models is an iterative process often requiring significant theoretical advancements. Simplified models, such as a toy model of a black hole in a lower dimension or with specific symmetries, can be used initially to make calculations tractable. The goal is to demonstrate core principles: the quantum nature of the singularity, the mechanism of information transfer, and the emergence of expanding spacetime. These simplified models, while not directly describing our universe, offer crucial insights and guide the development of more realistic, higher-dimensional models.

Effective field theories are crucial for constructing these models. In extreme gravitational environments, quantum effects can be treated as perturbations on a classical background or vice versa. However, at the Planck scale, where quantum gravity dominates, these approximations are often insufficient. Therefore, theories offering a more fundamental description of quantum gravity, such as string theory and loop quantum gravity, are essential.

Consider, for example, string theory's mathematical description of black hole evaporation. Calculations have shown that radiation emitted by these black holes can carry information about their interior. These calculations often map the black hole's complex gravitational dynamics to a simpler quantum field theory on its boundary. If a black hole is viewed as a collection of fundamental stringy degrees of freedom, Hawking radiation can be seen as the thermal emission of these entities, carrying their quantum information. The mathematical framework would then involve calculating the spectrum and correlations of this radiation, demonstrating that it is not purely thermal but possesses subtle quantum correlations encoding the black hole's history and, potentially, the initial state of a new universe.

Advanced mathematical techniques are vital for formulating these models. Differential geometry describes the curved spacetime of black holes and the emergent universe, while quantum field theory describes matter and energy in these spacetimes. Quantum information theory is crucial for understanding information encoding, processing, and transfer. Additionally, statistical mechanics is used to understand Hawking radiation's thermal properties and black hole entropy.

The exploration of emergent phenomena is central to achieving a unified mathematical description of the universe. Just as temperature and pressure are emergent properties of atomic behavior, gravity and spacetime might emerge from more fundamental quantum degrees of freedom. If our universe originated from a black hole, the laws of physics could be emergent properties of the parent black hole's quantum state at its boundary or within its interior. Mathematical models would then aim to identify these fundamental degrees of freedom and derive physical laws from their collective behavior. This is a profoundly challenging area requiring novel mathematical formulations to describe the transition from a quantum realm to a classical, spacetime-based reality.

Developing mathematical models for a universe emerging from a black hole highlights the scientific method's power to tackle seemingly intractable problems. It involves translating a significant astrophysical and cosmological hypothesis into mathematical language, subjecting it to rigorous analysis, and seeking testable predictions. This elegant process uncovers deep connections between disparate areas of physics: the extreme gravity of black holes, the quantum nature of information, and the cosmos's origin and evolution. While ultimate verification awaits further observational evidence and theoretical breakthroughs, ongoing work in constructing these mathematical frameworks continually pushes the boundaries of our cosmic understanding.

Precisely understanding quantum fluctuations near and within a black hole's horizon is crucial for mathematically describing such a transition. If we consider a black hole as a quantum system, its evaporation can be seen as a quantum mechanical process governed by the Schrödinger equation or its relativistic equivalent. The initial state would be the highly condensed matter forming the black hole, and the final state would be Hawking radiation, potentially with a remnant or transformed entity that initiates the new universe. The challenge lies in defining the Hilbert space of states for such a system and deriving the evolution operator governing its dynamics, making theories of quantum gravity indispensable for providing the fundamental rules of quantum mechanics in strong gravitational fields.

One mathematical modeling approach involves exploring black hole analogues in condensed matter systems. Superfluids or Bose-Einstein condensates can mimic black hole event horizons, such as sonic horizons. Studying the quantum behavior of excitations in these systems can offer insights into Hawking radiation and information scrambling in actual black holes. The mathematical equations describing these condensed matter analogues, typically well-understood quantum field theories, can be adapted and extended to the gravitational context. This allows physicists to test theoretical concepts in a laboratory setting, even with different underlying physics, as mathematical similarities in governing equations can reveal deep structural connections.

Quantum entanglement entropy is another critical aspect. As a black hole evaporates, entanglement between Hawking radiation and the remaining black hole's degrees of freedom must be accounted for. If information is preserved, the entanglement entropy of the emitted radiation should increase over time, allowing for information retrieval. Mathematical models tracking this entanglement entropy are vital for verifying unitarity. For a universe emerging from a black hole, its initial state would be intimately linked to the entanglement structure of the Hawking radiation. The mathematical description would need to show how these entangled states evolve to form the Big Bang's initial conditions.

The "firewall" concept at the event horizon, a potential resolution to the black hole information paradox, also requires careful mathematical formulation. If complementarity holds, and no firewall exists for an infalling observer, then entanglement between early and late Hawking radiation must be managed without violating quantum mechanics for an external observer. Mathematical models exploring these possibilities might involve new degrees of freedom or modifications to quantum field descriptions near the horizon. If a universe emerges from this process, the mathematical framework would need to explain how information is coherently transferred across this potentially complex spacetime region.

The holographic principle serves as a powerful guiding principle for constructing these mathematical models. If a black hole's interior physics is encoded on its event horizon, the mathematical description of the universe's origin from a black hole can be reformulated in terms of this boundary's dynamics. This suggests that the fundamental degrees of freedom responsible for the Big Bang might reside on a lower-dimensional surface, possibly the event horizon of a precursor black hole. The mathematical task then becomes describing the quantum field theory on this boundary and demonstrating its evolution into the three-dimensional expanding universe we observe. This approach leverages the successes of the AdS/CFT correspondence, a concrete realization of holography, and seeks to extend these ideas to more realistic cosmological scenarios.

The mathematical formulation of a universe born from a black hole is not merely theoretical abstraction; it is a crucial step in transforming a captivating idea into a scientifically viable hypothesis. Mathematical rigor provides the necessary tools to probe such a proposition's consequences, identify potential inconsistencies, and guide future research. It demands the development of new mathematical techniques and a deeper understanding of the interplay between gravity, quantum mechanics, and information. The ultimate goal is a coherent mathematical narrative that explains our universe's birth and resolves some of modern physics's most profound puzzles. This ongoing endeavor represents a frontier of scientific inquiry, where our knowledge boundaries are continuously redefined through mathematical reasoning and the relentless pursuit of understanding the cosmos.

The theoretical and mathematical foundations for a universe originating from a black hole have been established through prior research, encompassing general relativity, quantum mechanics, string theory, loop quantum gravity, and the crucial principle of information preservation. The focus now shifts from abstract equations to the practical application of computational exploration. Computer simulations are vital for testing the plausibility of this cosmic genesis. These are not mere thought experiments but sophisticated attempts to numerically replicate the extreme, yet hypothetical, conditions of a black hole's quantum evaporation and its subsequent transformation into a nascent cosmos.

Simulating the genesis of a universe from a black hole is an immensely complex computational undertaking, pushing current technological limits. At its core, the simulation must address the physics at the Planck scale, the smallest meaningful length scale in the universe, where both quantum mechanics and gravity are dominant. This is where our current understanding is limited, making numerical simulations an indispensable tool for exploring possibilities. The initial challenge lies in defining the simulation's starting conditions. If a black hole in a prior universe is envisioned undergoing evaporation, the simulation must model this process. This involves creating a discrete representation of spacetime, often a lattice or network, populated with fundamental degrees of freedom dictated by a chosen quantum gravity theory. For example, a loop quantum gravity-inspired simulation might represent spacetime with spin networks, with dynamics governed by quantum difference equations. These initial states would depict a highly compressed, quantum gravitational state within the black hole.

The parameters for these simulations are determined by the chosen theoretical model. A string theory-based model might involve tracking the interactions and dynamics of fundamental strings and branes in a high-dimensional spacetime. The "evaporation" phase would be simulated by these entities interacting, releasing energy and information, possibly as Hawking radiation. The key is to maintain the quantum coherence of this process, ensuring information is not lost but scrambled and encoded in the outgoing radiation. The simulation's output would be a sequence of quantum states representing the system's evolution. In a loop quantum gravity simulation, parameters would include the initial spin network configuration, the strength of quantum gravitational interactions, and the quantization of space and time. The simulation would then evolve these quantum states, step by step, to represent the theorized quantum bounce, which replaces a singularity.

The outputs of these simulations are incredibly rich but demand sophisticated analysis. Instead of a single, straightforward result, simulations often generate vast datasets that require interpretation. For a black hole evaporation scenario leading to a new universe, the simulation might track spacetime's state, the distribution of energy and matter, and the system's quantum entanglement properties. Early simulations might focus on simplified scenarios, such as a black hole in lower-dimensional spacetime or one with specific symmetries, to make computations manageable. These simplified models allow researchers to test underlying algorithms and gain intuition about quantum gravity's behavior in extreme regimes. For example, a simulation might focus on the event horizon's quantum state and its evolution as the black hole loses mass. The output could be a series of snapshots illustrating the changing quantum properties of this boundary, potentially revealing how information is preserved.

A crucial aspect that simulations aim to capture is the transition from the highly compact, possibly chaotic state within the black hole to the smooth, expanding spacetime of a new universe. This requires simulating the quantum bounce. In loop quantum gravity, a simulation might show a spin network collapsing to a state of maximum curvature and then rapidly expanding, with quantum fluctuations driving the transition. The output would be a visualization of how the discrete quantum spacetime transforms into a continuous manifold that then begins to expand. Parameters would include specific loop quantization techniques and the collapsing matter's initial quantum state. The simulation would then track this state's evolution through the bounce, observing how spacetime's geometry changes and how matter and energy are distributed during this critical phase.

The data generated by these simulations can be visualized in various ways. Direct visualization of spacetime at the Planck scale is impossible, so scientists often represent key quantities. For example, spacetime curvature could be visualized as a heatmap, with intense colors indicating extreme conditions near the supposed singularity or during the quantum bounce. The distribution of matter and energy might be shown as evolving density fields. Quantum entanglement between different parts of the system can also be visualized, perhaps as a network of connections, illustrating how information is shared and preserved. For a universe emerging from a black hole, the simulation might visualize the initial high-density, high-energy state as a compact region that then rapidly expands, with the outgoing Hawking radiation (if modeled) carrying away information and influencing the emergent spacetime.

Scientists utilize simulations to explore how black hole evaporation might lead to the formation of a new universe, particularly focusing on how a parent black hole imprints the initial conditions of an emergent cosmos. These simulations aim to reproduce observable features of our universe. For example, subtle quantum fluctuations in an evaporating black hole could be amplified during the "bounce" that creates a new universe, thus seeding the large-scale structure, like galaxy distribution. The simulations model how these initial quantum uncertainties translate into specific patterns of density fluctuations, comparing the output to cosmological observations such as the cosmic microwave background radiation to assess the model's viability.

The computational demands for such simulations are immense, requiring powerful supercomputers and optimized algorithms to resolve quantum gravity physics at incredibly small scales over vast cosmic times. Researchers often use numerical methods like finite difference or finite element to discretize and iteratively solve the equations of motion. Due to the current computational limitations in achieving Planck-scale resolution, approximations are frequently employed, focusing on crucial transition patches rather than the entire universe.

Another key area of simulation is the information paradox. If a new universe emerges from a black hole, information about what fell in must be encoded within the new universe or its emitted radiation. Simulations track information processing and transfer during evaporation and transformation, calculating quantities like entanglement entropy to demonstrate the conservation of total information. For instance, a simulation might show that as a black hole shrinks, the entanglement between the emergent spacetime and Hawking radiation precisely accounts for the original black hole's information content.

A crucial departure from classical general relativity in these simulations is the nature of the "singularity." Quantum gravity theories predict a region of high but finite curvature where spacetime "bounces" instead of an infinitely dense point. Simulations aim to capture this bounce, as seen in loop quantum gravity where spacetime discreteness at the Planck scale prevents infinite compression. Such a simulation would illustrate spacetime's quantum state reaching maximum curvature before rapid expansion, reflecting specific quantum geometry corrections from loop quantum gravity. The output could be a graph showing a smooth transition from contraction to expansion without a singularity.

Visualizations of these simulated cosmic births are often awe-inspiring, depicting intense light and energy rapidly expanding from a chaotic quantum state into an ordered classical universe. These are direct outputs of numerical solutions, showing a compressed spacetime "exploding" outwards, with nodes spreading to form new, expanding geometry, where color and intensity represent energy density or curvature.

The refinement of these simulations is an iterative process, starting with simplified models and gradually increasing complexity. Researchers validate their techniques against known physics or analytical results before tackling the full problem. For example, a simulation might first model Hawking radiation from a semi-classical black hole before incorporating full quantum gravity effects.

The parameters in these simulations are crucial and derived directly from theoretical models, such as string theory's compactification of extra dimensions or loop quantum gravity's holonomies and fluxes. The simulation output is highly sensitive to these parameters, requiring extensive exploration of the parameter space to find physically plausible results.

Simulations also allow for the exploration of alternative cosmological models. If our universe resulted from a black hole bounce, variations in initial conditions or quantum gravity effects could lead to different types of universes. Scientists can explore these variations to see if any match the observed properties of our own cosmos, for instance, by varying "dark energy"-like behavior to replicate observed accelerated expansion.

The ultimate goal is a predictive framework. If a simulation based on a specific quantum gravity theory can successfully reproduce observed features of our universe from a black hole-like initial state, it provides strong evidence for that theory and the black hole-to-universe hypothesis. This requires accurate simulations and robust methods to compare their outputs with astronomical observations, often indirectly translating simulation results into predictions for measurable quantities like the statistical properties of the cosmic microwave background or matter distribution.

Computational simulations serve as a vital link between theoretical concepts and observable reality. They enable us to explore the implications of our deepest cosmological ideas, guided by mathematical rigor and computational power. While direct observation of a black hole birthing a universe remains impossible, simulations offer a powerful avenue for scientific inquiry, pushing the boundaries of our understanding of gravity, quantum mechanics, and the fundamental nature of existence. The continuous advancement of computing power and simulation techniques promises to further illuminate this fascinating, speculative, yet scientifically grounded cosmic scenario. The visualizations produced from these simulations, though largely theoretical, provide a glimpse into the immense complexity and potential beauty of such a profound cosmic event, transforming abstract equations into dynamic, evolving cosmic landscapes. These simulations are indispensable tools in our quest to answer humanity's most fundamental questions about our origins.

As we've explored, the intricate interplay of mathematical models and sophisticated simulations translates the theoretical genesis of a universe from a black hole into a form that can be compared against the cosmos we inhabit. However, the most crucial and ultimately necessary step in this scientific endeavor is the rigorous comparison of these simulation outputs with the vast array of observational data collected by astronomers and cosmologists. This is the point where theoretical physics transitions from speculative exploration to empirical validation. The elegance of a simulated cosmic birth, regardless of its mathematical soundness or computational intensity, remains a fascinating hypothesis until it can convincingly replicate the universe we actually observe.

Key Observational Benchmarks for Validation:

1. **Expansion History of the Universe:** Cosmological observations, particularly from supernovae surveys and the cosmic microwave background (CMB), offer a remarkably detailed account of our universe's expansion over billions of years. Simulations of a black hole-born universe must generate a specific expansion rate at different epochs.
    - **Questions for Simulation Validation:** Does the simulated scale factor evolve consistently with the observed Hubble constant and its evolution? Do the parameters within the quantum gravity model, when evolved through the simulated bounce, naturally produce an inflationary period, followed by radiation-dominated and then matter-dominated eras, culminating in the currently observed accelerated expansion driven by dark energy?

- **Discrepancy Indication:** A significant difference in the predicted early universe deceleration rate compared to that inferred from CMB anisotropies would indicate a mismatch, requiring refinement of the simulation's initial conditions, the underlying physics of the bounce, or even the fundamental quantum gravity theory itself.

2. **Cosmic Microwave Background (CMB) Radiation:** The CMB, a faint afterglow of the Big Bang, provides a snapshot of the universe when it was approximately 380,000 years old. It is incredibly uniform but contains tiny temperature fluctuations (anisotropies) that seeded all large-scale structures we see today.

   - **Simulation Requirement:** Simulations of a black hole-to-universe scenario must reproduce the statistical properties of these anisotropies, including the power spectrum, which describes the amplitude of fluctuations as a function of angular scale. The specific peaks and troughs in the CMB power spectrum are highly sensitive to the universe's initial conditions, including the density of matter, dark energy, and the nature of primordial fluctuations.
   - **Successful Simulation Outcome:** A successful simulation would produce a universe whose nascent quantum fluctuations, amplified during the black hole bounce and subsequent ex-

pansion, closely match the observed CMB power spectrum.

- **Discrepancy Indication:** A deficit of power on large scales or an excess on small scales in the simulated power spectrum would signal a potential discrepancy requiring attention.
- **CMB Polarization (E-modes and B-modes):** CMB polarization patterns are also crucial. E-modes are sourced by density fluctuations, while B-modes are generated by gravitational waves from inflation or other exotic processes. Detecting primordial B-modes would strongly support a period of rapid, early inflation.
- **Simulation Requirement for Polarization:** A simulation must account for the generation of density fluctuations and, if the model includes an inflation-like mechanism, predict a specific level of primordial gravitational waves, which would manifest as B-mode polarization in the CMB.
- **Constraint on Models:** The absence or presence and amplitude of these B-modes provide powerful constraints on inflationary models and, by extension, on any model of cosmic genesis aiming to replicate our universe's observed features.

3. **Large-Scale Structure of the Universe:** The distribution of matter today, as mapped by large-scale structure surveys, offers another critical benchmark.

Galaxies are not randomly distributed but form a cosmic web of filaments, clusters, and voids, a direct consequence of gravitational amplification of initial density fluctuations (observed in the CMB).

- **Simulation Requirement:** Simulations of a black hole origin must evolve from their simulated early-universe conditions to produce a cosmic web that statistically matches these observations. This involves comparing the simulated power spectrum of matter distribution at later times with the observed galaxy correlation function.
- **Key Parameters:** Important parameters include the growth rate of structure (sensitive to the amount of matter and dark energy) and the bias of galaxies relative to the underlying dark matter distribution.

4. **Nature of Dark Matter and Dark Energy:** These constituents make up about 95% of the universe's energy density, presenting a significant challenge and opportunity for validation.

    - **Simulation Requirement for Dark Matter:** While simulations may not directly produce dark matter particles, they must accommodate their observed abundance and gravitational effects.

- **Simulation Requirement for Dark Energy:** If the model predicts a mechanism for the emergence of dark energy or a component that mimics its behavior, this must be quantitatively consistent with observations.
- **Alignment Requirement:** For instance, if the quantum gravity theory predicts a specific relationship between the initial conditions of the emergent universe and the effective cosmological constant driving late-time acceleration, this relationship must align with the observed value of dark energy.

The black hole-to-universe hypothesis suggests that our universe originated from the evaporation of a black hole in a previous cosmic epoch. This theory proposes that subtle anomalies in the Cosmic Microwave Background (CMB), unusual correlations in matter distribution, or specific patterns in gravitational waves could serve as observational signatures of this event. Some theories even posit a "bounce" within a black hole, where quantum processes might leave imprints on the emergent spacetime, potentially detectable if they survive the universe's expansion.

Validating such a cosmological model is an iterative and challenging process, adhering to the scientific method of hypothesis, prediction, observation, and refinement. However, direct experimentation is impossible, as our universe is the only instance we can observe. Therefore, the model's strength lies in its ability to explain a wide range of cosmological observations with minimal assumptions, all derived from a fundamental theory of quantum gravity. For example, if a quantum gravity framework, when simulated through a black hole bounce, can explain CMB anisotropies, large-scale structure, light element abundance, and accelerated expansion, it gains significant support. Conversely, if it requires ad hoc adjustments, its explanatory power diminishes.

Falsifiability is crucial; a scientific model must be disproven if its predictions are absent or if it fails to predict observed phenomena. If a quantum gravity simulation of a black hole bounce cannot reproduce observable features or predicts a fundamentally different universe, it would be falsified. Ongoing advancements in observational cosmology, with increasing precision in measurements, provide increasingly stringent tests for such models.

Translating simulation results into observable quantities often requires further theoretical work. For instance, evolving quantum states of spacetime and matter forward using classical general relativity is necessary to predict CMB temperature and polarization fluctuations, a process that involves assumptions about the transition from quantum gravity to classical spacetime.

Ultimately, validating a cosmological theory, including the black hole-to-universe hypothesis, depends on its ability to accurately predict observable phenomena confirmed by astronomical data. The expansion history, CMB patterns, galaxy distribution, and the nature of dark energy and dark matter are all critical touchstones. The success of any cosmological model is judged by its fidelity to the universe we observe, driving the continuous dialogue between theory and observation.

The Lambda-CDM model, while successful, faces significant challenges such as the nature of dark energy and dark matter, as well as the flatness, horizon, and initial singularity problems. These issues necessitate alternative models, among which is the black hole-to-universe hypothesis.

Other models, like Roger Penrose's Conformal Cyclic Cosmology (CCC), also address the initial singularity by positing an infinite sequence of cosmic expansions and contractions, with the Big Bang as a transitional phase. CCC suggests that Hawking radiation will eventually dominate, leading to a state that can be rescaled into a new inflationary epoch, offering a parallel to the black hole-to-universe hypothesis. However, the exact mechanisms for the "bounce" differ, with CCC relying on conformal transformations and Hawking radiation, while our hypothesis is grounded in quantum gravitational descriptions of a black hole's interior. Additionally, observational signatures vary, as CCC predicts specific concentric circles in CMB temperature anisotropies. Differentiating between these bounce mechanisms will require comparing their predictive power against precise CMB data.

Multiverse theories represent another broad class of alternative models, proposing that our observable universe is one of countless others, each with unique physical laws, constants, and initial conditions. These scenarios can arise from eternal inflation, string theory landscapes, or even quantum mechanics' Many-Worlds Interpretation. In eternal inflation, expanding spacetime constantly generates "bubble universes," including our own. If our universe originated from a black hole in a parent universe, that parent universe could itself be part of a larger multiverse. In this context, the black hole hypothesis could be viewed as a mechanism for generating new universes within an existing multiverse. A key challenge with multiverse theories is the difficulty in making direct, testable predictions for our specific universe. While they might explain the fine-tuned properties of our universe through the anthropic principle, empirically verifying other universes or their genesis processes is difficult. In contrast, the black hole origin hypothesis aims to explain our universe's observed properties through a specific physical process. Although it might eventually be integrated into a broader multiverse framework, its primary goal is to provide a testable mechanism for our universe's existence based on its observable features. The distinction is subtle but crucial: a multiverse offers a context, while the black hole hypothesis proposes a specific mechanism for genesis within any such context.

Modifications to the standard cosmological model (Lambda-CDM) also provide alternative explanations without necessarily invoking new universes or cycles. These modifications often address specific discrepancies or introduce new physics to resolve issues like dark energy, dark matter, or the flatness and horizon problems. For example, modified gravity theories alter Einstein's general relativity on cosmic scales, potentially explaining accelerated expansion without a cosmological constant or dark energy. Similarly, alternative dark matter candidates or adjustments to particle physics could solve the missing mass problem. It is plausible that the quantum gravity theory describing the black hole bounce might inherently incorporate features resembling modified gravity or naturally explain phenomena attributed to dark matter or dark energy. The exotic states of matter and spacetime during a quantum bounce could, upon expansion, lead to effective gravitational behavior that deviates from standard general relativity. Alternatively, the emergence from a black hole might imbue the nascent universe with specific quantum fields or vacuum energy states that mimic dark energy. However, a critical difference is that our hypothesis seeks a unified origin story, potentially explaining the universe's properties from a single, complex theoretical framework, whereas Lambda-CDM modifications often tackle specific problems in isolation or by introducing new, independent components. The challenge for our hypothesis is to demonstrate that the physics of a black hole bounce not only creates a universe but also naturally produces the correct expansion history, structure formation, and cosmic constituents that align with or explain observations currently attributed to dark energy and dark matter, rather than simply adding new parameters or modifications to the standard model.

The black hole universe hypothesis offers compelling alternatives to existing cosmological frameworks like the Lambda-CDM model, cyclic models, multiverse theories, and ad hoc modifications to Lambda-CDM.

Critiques of Existing Models and Solutions Offered by the Black Hole Universe Hypothesis:

- **Lambda-CDM Model:**
    - **Weaknesses:** Suffers from "fine-tuning" problems, where cosmological parameters seem arbitrarily set for a complex universe, and the initial Big Bang singularity poses a conceptual challenge, breaking down current physics.
    - **Black Hole Universe Solution:** A quantum gravity-governed transition through a black hole singularity could naturally establish initial expansion conditions, avoiding the extreme fine-tuning. This "bounce" could also impart specific initial fluctuations, resolving the singularity problem inherently.
- **Cyclic Models (e.g., Conformal Cyclic Cosmology - CCC):**

- **Strengths:** Resolve the initial singularity and address the "what came before?" question.
- **Weaknesses:** Observational challenges persist, such as definitively detecting predicted concentric circles in the Cosmic Microwave Background (CMB) for CCC.
- **Black Hole Universe Solution:** Predicts distinct observational signatures, including imprints from the black hole interior on the CMB or large-scale structure, or unique gravitational wave signals. Its success hinges on the confirmation of these signatures. If CCC's predictive power for CMB features falls short, and the black hole bounce model consistently explains a broader range of cosmological data (expansion history, large-scale structure, CMB anisotropies) through quantum gravity-derived parameters, it gains an advantage.

- **Multiverse Theories:**

  - **Weaknesses:** Face a significant hurdle in testability; empirical verification of other universes is lacking, keeping them in the realm of philosophical speculation.
  - **Black Hole Universe Solution:** Aims to be a scientific theory about *our* universe, with validation or refutation based on its ability to accurately describe the cosmos we inhabit. If it predicts specific, observable consequences for

our universe, distinct from standard cosmology or other alternatives, and these are confirmed, it provides strong evidence for its validity, independent of the multiverse concept. The multiverse might explain *why* a black hole bounce could occur, but the hypothesis itself requires validation from observations within our universe.

- **Modifications to Lambda-CDM:**
  - **Weaknesses:** Often ad hoc fixes for specific problems (e.g., new scalar fields for inflation or specific forms of dark energy) without offering deeper, unifying explanations.
  - **Black Hole Universe Solution:** If successful, it would offer a more fundamental and unified explanation for multiple cosmological phenomena. If quantum gravity's description of a black hole bounce naturally leads to inflation, sets initial conditions for structure formation, and even dictates the nature or abundance of dark matter and dark energy, it would be a significant advancement over piecemeal modifications.

Explanatory Depth and Parsimony:

The core comparison revolves around explanatory depth and parsimony. Does the black hole origin model, with its quantum gravity and black hole assumptions, explain more observed features of our universe with fewer arbitrary parameters than models that simply add components to standard cosmology?

Initial Conditions:

- **Standard Cosmology:** Posits finely-tuned initial conditions at the Big Bang singularity.
- **Inflationary Cosmology:** Proposes a mechanism to smooth inhomogeneities and generate structure seeds.
- **Black Hole Universe Hypothesis:** Suggests that the quantum gravitational state within a black hole, or its resolution process, naturally sets these initial conditions. The quantum fluctuations that become CMB anisotropies and galaxy seeds would originate from the bounce's quantum processes. This offers a potentially more fundamental origin story for these crucial cosmological features, contrasting with standard inflation, which often introduces parameters (like spectral index and tensor-to-scalar ratio) not fully explained by the underlying theory.

The nature of dark energy is another crucial point of comparison. If the quantum gravity framework, which describes the black hole bounce, naturally predicts a vacuum energy density or scalar field behavior mimicking dark energy at later stages, this would strongly support the model's explanatory power. Dark energy currently poses one of cosmology's greatest mysteries, with the cosmological constant (Λ) presenting an immense fine-tuning problem. If a black hole origin scenario can resolve this mystery through its fundamental physics, it would offer a significant advantage over Lambda-CDM. Similarly, insights into the particle nature or distribution of dark matter would further strengthen its position.

The potential for unique observational signatures is a key differentiator. While standard cosmology has clear predictions for the CMB power spectrum and large-scale structure, and cyclic models forecast specific patterns, the black hole universe hypothesis might yield entirely different or subtle signatures. These could include deviations from standard Lambda-CDM predictions in small-scale structure formation, specific correlations in galaxy and dark matter distribution reflecting initial conditions from the bounce, or even unique gravitational wave backgrounds differing from those predicted by standard inflationary models. For example, a bounce might generate a spectrum of gravitational waves with distinct frequency dependence or polarization properties compared to inflationary gravitational waves. Detecting such signatures would directly test the black hole origin hypothesis and help distinguish it from other cosmological paradigms.

The robustness of the underlying theory is also vital. The black hole universe hypothesis is inherently linked to developing a complete theory of quantum gravity, such as string theory or loop quantum gravity. If progress in these fundamental theories provides a consistent description of black hole interiors that naturally leads to a bounce and subsequent expansion, it would significantly support this cosmological hypothesis. In contrast, modifications to Lambda-CDM, while possibly phenomenologically successful, may not be rooted in a similarly fundamental theoretical framework.

In summary, comparing the black hole universe hypothesis with other cosmological models reveals a diverse landscape of ideas, each attempting to answer fundamental questions about cosmic origin and evolution. While standard Lambda-CDM remains the benchmark, its fine-tuning and singularity problems motivate alternatives. Cyclic models offer a way to bypass the singularity but face their own observational challenges. Multiverse theories provide a grander context but struggle with testability. Modifications to Lambda-CDM address specific issues but may lack fundamental depth. The black hole universe hypothesis presents a compelling vision of cosmic genesis, potentially resolving singularity and fine-tuning problems, and offering a unified explanation for various cosmological phenomena. Its ultimate success depends on its ability to generate unique and verifiable observational signatures that can be tested against the vast and ever-growing body of astronomical data, thereby carving its own niche within the grand narrative of cosmology. This comparative analysis highlights that the quest for understanding our universe is an ongoing dialogue between theoretical innovation and empirical verification, with each model contributing to the larger picture of cosmic inquiry.

The enduring quest to reconcile the enigmatic realm of quantum mechanics (the very small) with general relativity's description of macroscopic spacetime is the foundation for future advancements in cosmological modeling. For a hypothesis as deeply connected to the fundamental nature of gravity and spacetime as the emergence of a universe from a black hole, the refinement and validation of its underlying mathematical framework are inextricably linked to progress in quantum gravity. This subsection explores the pivotal role anticipated theoretical breakthroughs will play in shaping our understanding and potentially confirming or refuting such a cosmology. The pursuit of a unified theory of everything, a single framework encompassing all fundamental forces and particles, stands as the ultimate arbiter of many currently speculative cosmological ideas, including the black hole universe hypothesis.

Current mathematical models describing a black hole bounce and the subsequent inflationary expansion of a new universe are incomplete. They often rely on extrapolations of known physics into regimes where it is known to break down, especially at the extreme densities and curvatures of the Planck epoch. For instance, the resolution of the singularity within a black hole often uses simplified models or approximations of quantum gravitational effects. A truly complete theory of quantum gravity, such as loop quantum gravity or string theory, when fully developed, is expected to provide a more rigorous and detailed description of these Planck-scale phenomena. This could lead to a recalculation and refinement of parameters governing the bounce, such as the precise minimum scale achieved before expansion, the nature of the quantum state at that point, and the spectrum of primordial fluctuations generated. Such refinements are crucial. If a fully realized theory of quantum gravity, for example, predicts a specific set of initial conditions for an expanding universe that directly matches the observed anisotropies in the Cosmic Microwave Background (CMB) with unprecedented accuracy, it would offer powerful validation for the black hole bounce mechanism as the origin of our universe. Conversely, if predictions derived from the mature quantum gravity theory significantly deviate from cosmological observations, it would necessitate a re-evaluation or even abandonment of the black hole origin hypothesis.

Future theoretical advancements in quantum gravity offer a profound impact on the black hole universe hypothesis, particularly in understanding spacetime, dark energy, and dark matter. Theories like loop quantum gravity and string theory propose a quantized or string-based nature of spacetime, respectively. This granular view could redefine singularities, suggesting non-classical regions or "fuzzball" structures, and allow for a more precise mathematical description of the transition from a quantum-gravitational state to an expanding universe. Such descriptions might reveal observable signatures in the early universe, such as CMB polarization or gravitational wave signals.

A deeper understanding of quantum gravity could also provide fundamental explanations for dark energy and dark matter, which are currently treated as fine-tuned parameters in cosmological models. It's conceivable that quantum gravitational processes within a black hole bounce could naturally generate the observed energy densities of dark energy or the constituents of dark matter, eliminating the need for additional fine-tuning. Successfully demonstrating that these cosmological mysteries are emergent properties of a black hole-born universe would be a significant triumph for the hypothesis.

Beyond refining existing models, quantum gravity has the potential to unlock entirely new avenues of inquiry, revealing previously unconceived phenomena related to black hole interiors and their role in a universe's emergence. For instance, insights from quantum entanglement and black hole thermodynamics, particularly concerning the information paradox, could inform how a universe is encoded or released from such an object. This dynamic interplay between theoretical understanding of black holes and the origin of the universe ensures that progress in one area will fuel the other.

Finally, the advancement of computational cosmology is heavily reliant on these theoretical breakthroughs. As our mathematical understanding of quantum gravity and the black hole bounce grows, so too will our ability to perform high-fidelity numerical simulations. These simulations can rigorously test theoretical predictions against observational data, such as comparing predicted gravitational wave spectra or structure formation patterns with actual cosmological observations. The ability to simulate complex, non-linear quantum gravitational processes will be a vital link between abstract theory and empirical verification, with future advancements in computational power enabling simulations that capture the full richness of a universe emerging from a quantum-gravitational transition.

The mathematical models of a black hole universe require refinement through testable predictions that differentiate them from other cosmological models. Future theoretical advancements are crucial for guiding observational efforts. For instance, if quantum gravity predicts a unique imprint on the cosmic microwave background (CMB) polarization during the transition from a black hole interior to an expanding universe, or distinct non-Gaussianities in primordial density fluctuations, or a unique correlation in large-scale galaxy distribution, these would become key targets for observational cosmology. The continuous refinement of these models is not merely a theoretical exercise; it's essential for identifying evidence that could either confirm or invalidate the hypothesis. Without clear, distinguishing predictions from robust theoretical frameworks, even the most elegant mathematical models remain speculative.

The evolution of theoretical physics will also necessitate the development of new mathematical formalisms. Current mathematical tools may be inadequate to fully capture the complexities of a quantum gravitational bounce. Future theoretical insights might introduce novel mathematical languages or concepts, possibly from areas such as category theory, noncommutative geometry, or advanced topological methods, that are better suited for describing the transition from a singularity-resolved black hole interior to an expanding spacetime. Such frameworks could provide new ways to understand initial conditions, the mechanism of inflation, and the generation of primordial structure, highlighting the iterative nature of scientific progress.

Consider the implications for inflationary cosmology, a cornerstone of the standard Lambda-CDM model. While the black hole bounce hypothesis can accommodate inflation, a more developed quantum gravity theory might offer a more fundamental explanation for inflation or alternative scenarios for how a universe could enter an inflationary trajectory after a bounce. For example, if the quantum state at the bounce point is highly energetic and unstable, it might naturally evolve into an inflationary phase without the need for separate, finely-tuned inflaton fields. If future theoretical work demonstrates that a black hole bounce naturally leads to the specific type and duration of inflation required to solve the horizon and flatness problems, it would significantly strengthen the hypothesis, offering a more unified and elegant genesis story.

The early conditions established by a black hole bounce could significantly influence the mathematical modeling of the late-time universe, including dark energy behavior and the formation of large-scale structures. Should the nascent universe retain specific non-gravitational fields or properties from its quantum gravitational origin, these could lead to long-term consequences that deviate from conventional cosmological predictions. For example, residual quantum entanglement from the bounce might manifest as subtle correlations in matter distribution across vast cosmic scales or alter the precise rate of cosmic acceleration in ways not anticipated by a simple cosmological constant. Further theoretical work is necessary to trace these potential consequences throughout cosmic history, from the Planck era to the present, to identify distinct observational signatures.

As our understanding of quantum gravity deepens, the very definition of a "black hole" may need re-evaluation. If, as some theories propose, black holes are not classical singular points surrounded by event horizons but rather complex, quantum-gravitational objects without true singularities, then the mathematical description of a universe emerging from such an object would require adaptation. This could present a richer and more nuanced picture of cosmic origins, potentially involving the interaction of multiple quantum entities or the emergence of spacetime from a more intricate quantum substrate. Mathematical models would need to evolve in tandem with these fundamental shifts in our comprehension of black hole physics.

Ultimately, future theoretical advancements are crucial for developing a more complete, self-consistent, and predictive mathematical framework for the black hole universe hypothesis. This involves not only addressing the current limitations of our models but also exploring new theoretical landscapes that may arise from the ongoing quest for a theory of quantum gravity. The process will entail refinement, validation, and potentially, radical re-imagination. As our understanding of the fundamental laws of physics deepens, so too will our capacity to model the most profound questions about our cosmic origins, moving the hypothesis from speculative possibility towards empirically verifiable scientific inquiry. The success of the black hole universe hypothesis will largely be a testament to the power and elegance of the future mathematical and theoretical physics that underpins it.

CHAPTER 9

Philosophical and Existential Implications

The long-standing human inquiry into our cosmic place, from ancient stargazers to modern astrophysicists, faces a profound challenge: the hypothesis that our universe exists within a black hole. This isn't just a change of address; it's a fundamental redefinition of existence itself. If our universe is indeed the interior of a black hole in a parent universe, it dramatically alters our understanding of scale, situating our entire observable reality within a vaster and potentially incomprehensible cosmic framework.

Consider the audacious implications: our Milky Way, Andromeda, the vast cosmic web, billions of galaxies, and trillions of stars—all confined within the gravitational embrace of an entity that, from the parent universe's perspective, might appear as a point of immense density or a complex quantum object. This would demote our cosmic neighborhood, and everything we observe, to a mere substructure. It forces us to confront the idea that our perceived cosmic horizon is, in fact, an internal membrane within a larger, unseen entity. The impact on our self-perception is staggering; we could be likened to inhabitants of a water droplet in a vast ocean, oblivious to its true extent or the life beyond our contained environment.

This perspective also compels us to reconsider the concept of cosmic uniqueness. Humanity has long pondered extraterrestrial life. The black hole universe hypothesis adds a deeper layer to this question. If our universe is one of potentially countless universes born from black holes in a parent cosmos, the idea of our universe being singular or special dissolves. We become one instance among many, a specific manifestation of a cosmic process. This doesn't diminish the wonder of our existence but situates it within a potentially infinite fractal of realities. The search for extraterrestrial life within our universe would then be complemented, or perhaps superseded, by the question of "extradimensional" or "extra-universal" life—beings or processes inhabiting the parent universe or other universes born from its black holes.

Philosophical ramifications extend to our understanding of origins. In standard cosmology, the Big Bang is a singular beginning. While quantum cosmology offers explanations for its conditions, the ultimate "why" of the initial singularity remains a mystery. The black hole universe hypothesis, however, offers a potential genealogical link. Our Big Bang becomes a "bounce" or quantum transition within a pre-existing black hole. This shifts the question from "What happened at the beginning?" to "What was the parent black hole, and what was its origin?" While not eliminating the need for a prime mover, it offers a more concrete framework for tracing cosmic ancestry, suggesting a cyclical or nested nature where universes are born from the demise of others in a grand lineage.

The scientific method undergoes a subtle transformation. Our ability to observe is limited by the speed of light and the universe's age. If we exist within a black hole, these constraints might signify not merely physical boundaries, but fundamental differences between our internal universe and the external reality of the parent cosmos. The laws of physics, highly effective within our universe, could be a subset of a broader set governing the larger structure. This introduces the compelling, yet potentially frustrating, notion that certain core questions about our ultimate origins may forever elude empirical investigation, accessible only through theoretical conjecture or indirect deduction. It demands scientific humility, acknowledging that our current understanding may represent a localized description within a more universal reality.

The long-held human sentiment of cosmic isolation could be reinterpreted. While the standard model positions us at a unique point in cosmic history, stemming from a singular event, the black hole universe hypothesis could foster a sense of cosmic interconnectedness, albeit in an profoundly alien manner. If black holes are prevalent in the parent universe, and if each contains a nascent universe, then our universe is simply one among countless cosmic nurseries. This implies a shared ancestry, a common genesis, linking our reality to innumerable others. The universe ceases to be an isolated entity and instead becomes a thread woven into a far grander, more intricate tapestry. This could cultivate a sense of shared destiny with other potential universes, even if they remain eternally beyond direct observation.

The concept of finality is also called into question. In some philosophical interpretations of cosmology, the ultimate fate of our universe—whether it expands indefinitely, collapses in a Big Crunch, or succumbs to some other entropic end—represents a definitive conclusion. However, within the black hole universe framework, the end of our universe might merely be the precursor to the formation of a new black hole, which in turn could be the progenitor of another universe. This introduces a profound sense of cosmic recursion, an unending cycle of creation and transformation that transcends the lifespan of any single universe. It suggests that "endings" are simply transitions in a ceaseless, cosmic metamorphosis. The death of a star, the evaporation of a black hole, or the final heat death of our universe could all be stages in a much more expansive, ongoing cosmic narrative.

Our understanding of "real" or "fundamental" reality may also require recalibration. If our universe is an emergent phenomenon within a black hole, then the "reality" of the parent universe, with its own laws and constituents, could be considered more fundamental. Our physical laws, our particles, and our very existence would then be contingent properties, emergent features of a deeper, more encompassing reality. This aligns with certain philosophical traditions that posit a hierarchy of realities or modes of being. It prompts us to consider whether there are multiple levels of existence, with our universe occupying a specific, perhaps even transient, stratum. The pursuit of our ultimate nature would then involve investigating the properties of the black hole that birthed us, and perhaps, the nature of the spacetime in which that black hole exists.

The implications for our understanding of time are also substantial. Within a black hole, time as we perceive it is distorted by extreme gravity. If our universe originated from such a region, it raises questions about the initial state of time itself. Was time a property of the parent universe that was somehow transferred or transformed during the black hole bounce? Or did time, as we perceive it, only begin to exist with our universe? This could lead to a more nuanced perspective on time, not as a linear, absolute progression, but as an emergent property tied to the specific conditions of spacetime, potentially exhibiting different characteristics in different cosmic epochs or different universes. The "arrow of time" might be an internal characteristic of our universe, rather than a universal constant.

The theory that our universe exists within a black hole radically alters our understanding of reality. We typically define reality as the totality of everything that exists. However, if our universe is an interior, a contained cosmos, then reality must encompass a larger, external context. This raises a crucial philosophical question: what is the nature of this "outside"? Does it operate under the same fundamental laws as our own, or is it something entirely different, perhaps beyond our current comprehension?

Consider the analogy of a dream. While dreaming, the world within it feels completely real. We interact with its environments, characters, and events with a sense of authenticity. Yet, upon waking, we recognize that the dream reality was a mental construct. The "reality" of the waking world, with its independent existence and physical laws, is arguably more fundamental. Similarly, if our universe is inside a black hole, the physical laws we observe—the forces, particles, and spacetime—might be akin to the rules of a dream. They govern our experience but could be emergent properties, contingent on the specific conditions of our contained cosmos, rather than absolute, universal truths.

In this scenario, the "outside" would be the parent universe or the larger cosmic structure housing the black hole. Describing this outer reality challenges our imagination, as our current physics is intrinsically linked to our universe's spacetime and conditions. However, we can speculate by extrapolating from our understanding of black holes. In our universe, black holes are characterized by event horizons (points of no return) and singularities (points of infinite density, or regions of extreme but perhaps not infinite density in a quantum gravity context). If our universe lies beyond the event horizon of a black hole in another universe, then the parent universe would be the "normal" spacetime from which that black hole formed.

The physical laws in the parent universe might be similar, leading to the formation of black holes that then cradle new universes. Alternatively, the parent universe's laws could be vastly different, perhaps involving a completely different set of fundamental forces, elementary particles, or even a different number of spatial dimensions. The nature of the singularity within the black hole, which acts as the boundary between our universe and its origin, is a critical area of theoretical investigation. If it's a point-like singularity, our universe might have emerged from an infinitely dense point. If quantum gravity smooths out the singularity, then our universe could have begun from a region of exceptionally high but finite density and curvature, effectively a "bounce" from a previous cosmic epoch within the black hole.

This perspective also offers a potential reinterpretation of cosmic fine-tuning. The universe appears remarkably well-suited for the emergence and evolution of life, often discussed in terms of the "anthropic principle." If our universe is one of many born from black holes, and if the properties of these universes vary, then it becomes less surprising that at least one, our own, possesses the specific conditions necessary for life. We simply find ourselves in a universe conducive to our existence, a selection effect operating on a meta-universal scale. This doesn't diminish the wonder of our existence but provides a potential statistical explanation for fine-tuning, situating us within a probabilistic landscape of universes.

The very definition of "cosmology" might need to expand. Traditionally, cosmology has been the study of our universe—its origin, evolution, and structure. If our universe is embedded within a larger cosmic context, a complete cosmology would necessarily incorporate the study of these parent universes, their properties, and the mechanisms by which they spawn new realities. This suggests a multi-layered approach to cosmic study, where our local universe is just one chapter in a much grander, ongoing cosmic narrative. It would necessitate developing theoretical and observational tools not just to study the cosmic microwave background or the distribution of galaxies, but to potentially glean information about the cosmic environment from which our universe sprang.

The philosophical concept of "meaning" is also deeply affected. If our universe is not a unique creation but one of many, and if our existence is tied to the internal processes of a black hole in another realm, does this diminish or enhance the meaning of our lives? Some might find it a humbling perspective, reducing our perceived significance. Others might find it a source of awe and wonder, connecting us to a vaster, more intricate cosmic tapestry than ever imagined. The search for meaning might shift from seeking a singular purpose assigned by a creator to understanding our role and significance within a complex, generative cosmic system. Our individual lives and collective history would then be part of an even larger, unfolding story, written in the fabric of spacetime across multiple cosmic generations.

Ultimately, the hypothesis that our universe resides within a black hole compels us to confront the limits of our perception and the vastness of reality. It challenges anthropocentric views and invites a more humble, yet potentially more profound, understanding of our cosmic home. It suggests that the boundaries of our knowledge are not necessarily the boundaries of existence, and that our universe, for all its immensity, might be but a single, albeit wondrous, detail in an unimaginably larger cosmic artwork. This intellectual journey, from the comforting familiarity of our cosmic neighborhood to the startling implications of being a universe within a black hole, is a testament to the enduring human drive to understand our place, not just on Earth, or in our galaxy, but within the deepest mysteries of existence itself.

The notion of nested realities, where universes exist within other universes, prompts fundamental questions about the very fabric of existence. If our universe resides within a black hole, which in turn exists within a larger universe, does this hierarchy extend infinitely, a cosmic fractal of nested realities? Or is there a singular, foundational reality from which all others stem? This inquiry echoes ancient philosophical debates concerning monism and pluralism—whether reality is ultimately unified or composed of distinct entities.

Such a layered reality has profound philosophical implications. It challenges our inherent, often human-centric, belief that our experiences and observations encompass the entirety of existence. Our cosmic horizon, typically seen as a physical limit set by the speed of light and the universe's age, could instead be an internal boundary within a grander cosmic structure. This would render entities and phenomena "outside" our universe, in the parent cosmos, fundamentally beyond our direct observation. Our scientific understanding, dependent on empirical data from within our universe, might thus be inherently limited, describing only one layer of reality.

This leads to the concept of epistemic limits—the boundaries of what we can know. If our universe is a self-contained system, then questions about the parent universe's fundamental nature or the cosmic hierarchy's ultimate origin might lie beyond empirical science. We might be confined to philosophical speculation or theoretical inference, akin to a character in a novel unable to access the author's world. This doesn't invalidate our scientific endeavors but contextualizes them, highlighting the potential for a reality that surpasses our current observational and theoretical capabilities.

Consider the implications for our search for meaning and purpose. If our universe isn't the ultimate reality but a component of a larger cosmic structure, how does this impact our sense of significance? Some might find this perspective humbling, even diminishing, seeing ourselves as a minuscule part of an unfathomably vast and intricate whole. Others might find it exhilarating, a source of profound awe, connecting us to a cosmic narrative far grander than previously imagined. The quest for meaning might shift from seeking a singular, imposed purpose to understanding our role within a dynamic, generative system where new universes are continuously born. In this view, our existence could be part of a continuous cosmic lineage.

The concept of "boundaries" itself becomes fluid. We perceive the edge of the universe, the cosmic microwave background radiation, as the furthest we can see. If we exist within a black hole, this boundary is not an external edge but an internal one, a consequence of spacetime curvature. The black hole's event horizon, containing our universe, would be the true boundary separating our internal reality from the parent universe's external reality. But what is the nature of that boundary—a smooth transition, a quantum veil, or something more abrupt?

The very definition of "existence" may need re-evaluation. If our universe is an emergent phenomenon, its existence is contingent on the parent universe and the specific conditions that formed the black hole. This suggests a tiered ontology, where different levels of reality possess varying degrees of fundamentality. The parent universe might be considered "more real" or "more fundamental" than our own, simply by virtue of being the containing structure. This echoes philosophical ideas where the perceived world is a manifestation of a deeper, underlying reality.

The implications for time are also significant. Within a black hole, time behaves dramatically differently. As one approaches the singularity, extreme gravity distorts time. If our universe emerged from such a region, it raises questions about time's initial state. Did time as we know it begin with our universe, or was it a pre-existing property of the parent universe, carried over and transformed? The "arrow of time," our perceived unidirectional flow, might be an internal characteristic of our universe, a consequence of its specific thermodynamic and cosmological conditions, rather than an absolute, universal constant.

Furthermore, the concept of "emptiness" or "void" gains new dimensions. We often think of space as mostly empty, with matter sparsely distributed. However, if our universe is a black hole's interior, the "emptiness" we perceive could result from extreme gravitational conditions sweeping matter towards the singularity. The parent universe, from which the black hole originated, might be teeming with matter, energy, and phenomena vastly different from what we observe. What we consider the vacuum of space might, from the parent universe's perspective, be a region of immense activity or even a precursor to new cosmic structures.

The symmetry between inside and outside—our universe and the parent universe—is a fascinating area of contemplation. If our universe is a black hole interior, perhaps the parent universe, to its inhabitants, could also be the interior of an even larger structure. This could lead to an infinite regress of nested realities, a cosmic onion with endless layers, or there might be a foundational, ultimate reality at the "bottom" of this hierarchy. Understanding the "outside" would involve comprehending the process by which black holes form and potentially spawn new universes, a process likely governed by quantum gravity effects currently beyond our empirical reach.

The philosophical implications also extend to our understanding of consciousness and sentience. If our reality is a contained phenomenon, what does this suggest about consciousness? Is it an emergent property tied to our universe's specific physical laws and structures, or does it have a more fundamental existence that could transcend our cosmic boundaries? Could consciousness exist in the parent universe, observing or interacting with the black hole containing our reality in ways we cannot comprehend? These questions delve into the deepest mysteries of mind and matter, and the black hole universe hypothesis adds another layer of complexity to these age-old inquiries.

The very notion of "discovery" is also recontextualized. For centuries, science has explored the known and pushed the boundaries of ignorance, extending our reach into the unknown cosmos. If our universe is a contained system, "discovery" might also involve delving into our reality's internal structure, seeking to understand the precise nature of the black hole boundary and the conditions that led to our universe's formation. It might also involve developing entirely new theoretical frameworks, perhaps inspired by quantum gravity, that can bridge the gap between our internal physics and the potential physics of the external cosmos.

The idea of "other universes" extends beyond theoretical concepts like those found in inflation or string theory. If our universe is indeed the interior of a black hole, then these "other universes" could be the numerous black holes within the parent cosmos. Each such black hole might contain its own unique universe, complete with distinct physical laws and evolutionary histories. This perspective introduces a more fundamental distinction between the "observable universe" and the "total universe" than the limitation imposed by the speed of light, suggesting a cosmic landscape filled with entire universes rather than just galaxies and stars.

This viewpoint also challenges our intuitive understanding of space, time, and causality, which are based on our experiences within this universe. Applying these intuitions to the "outside" of our universe or to the black hole boundary can lead to paradoxes. For instance, the concepts of "before" and "after" might be different or even undefined in relation to the parent universe. Understanding the temporal connection and the mechanisms of emergence between our universe and a parent black hole is crucial for comprehending the nature of reality.

Ultimately, the hypothesis that our universe resides within a black hole necessitates a radical re-evaluation of our understanding of "reality" and the extent of our knowledge. It suggests a more intricate and interconnected tapestry of existence. Our universe's "inside" could be a dynamic cosmos or a meticulously crafted illusion, a temporary manifestation within a vaster, possibly eternal, cosmic structure. This intellectual journey, which questions the very foundations and boundaries of reality, highlights humanity's continuous pursuit of understanding and pushes the limits of our perception and philosophical comprehension into the deepest cosmic mysteries.

Considering our universe as being within a black hole in a larger cosmos also prompts a re-examination of the age-old philosophical debate between determinism and free will, seen through a new lens of cosmic causality. If our universe's genesis and evolution are inherently tied to the physical processes of a parent black hole, the concept of an uncaused cause or a truly spontaneous decision becomes uncertain. Are the events we observe, from galactic movements to subatomic interactions, merely the inevitable unfolding of a cosmic script dictated by the physics of the host universe and the specific conditions of the black hole that birthed us?

Causality typically implies a linear progression of events where actions lead to reactions. Free will often assumes the genuine ability to choose among multiple futures, independent of prior states. However, if our universe is an emergent property of a black hole's spacetime geometry, the causes shaping our reality might originate beyond our observable horizon. The fundamental laws of physics—like the strength of gravity, particle masses, and the fine-structure constant—could be determined by the conditions of the parent universe and the black hole's formation.

If these fundamental parameters are fixed by external, pre-existing conditions, then our universe's subsequent evolution, including the emergence of life and consciousness, might be a deterministic cascade. For example, if the mass and charge of an electron in our universe are not arbitrary but determined by the star that collapsed to form our containing black hole, then our physical reality is built upon a preordained foundation. This suggests that the "choices" available to the universe and its conscious inhabitants are constrained by this inherited blueprint.

This perspective doesn't necessarily negate the subjective experience of free will. Our feeling of making decisions is a powerful aspect of consciousness. However, from an external observer's or a deterministic cosmological viewpoint, these decisions could be seen as complex but predictable outputs of intricate physical and biological processes. The neural firings, influenced by genetics, environment, and past experiences, might meticulously trace a trajectory that appears as free choice to the individual but is, in fact, the inevitable result of preceding states.

The notion of cosmic causality extending beyond our observable universe poses a significant challenge to our understanding of agency. If the initial conditions that set our universe in motion were determined by events in the parent universe, then the entire history of our cosmos, from its emergence from the black hole to the present, could be considered a single, unfolding event without genuine alternative paths. This would imply that every quantum fluctuation, stellar formation, and evolutionary step on Earth was, in principle, predictable given complete knowledge of the parent universe and the black hole's internal dynamics.

The implications for human responsibility and morality are profound. If our actions are ultimately predetermined by a causal chain originating from events in a universe we cannot perceive, can we truly be held responsible for our deeds? Philosophers have debated this for centuries, exploring compatibilism (where free will and determinism coexist) and incompatibilism (where they don't). The black hole universe hypothesis adds a cosmic dimension to this debate, suggesting that the "determinism" is not just local or biological, but fundamental and cosmological.

Our current understanding of black holes and the singularity (or its quantum gravity equivalent) remains largely speculative, particularly concerning their interior physics. It's plausible that processes within black holes aren't purely deterministic in the classical sense. Given the inherent randomness of quantum mechanics, if our universe emerged from a black hole through a quantum process, an element of indeterminacy might exist at its very origin.

For example, the specific characteristics of our universe—such as fundamental constants and the initial distribution of matter and energy—could stem from unpredictable quantum fluctuations within a black hole. If such quantum events determined our universe's initial state, then diverse universes with varying physical laws might have arisen from similar black holes, introducing cosmic chance and potentially a form of cosmic free will or a less rigidly determined universe.

The concept of "causal loops" also becomes relevant. If our universe resides within a black hole, and that black hole's formation was influenced by a parent universe, which in turn is affected by its own black holes, we might be observing a complex, interconnected system where cause and effect aren't always linear. Could our universe subtly influence its parent? While counter-intuitive, certain theoretical frameworks in cyclic cosmologies and extreme gravitational environments don't entirely dismiss such abstract possibilities.

Furthermore, "free will" might require re-evaluation in the context of cosmic causality. Perhaps free will isn't the absence of causes, but rather the nature of those causes. If the causes originate within our universe from the intricate interplay of our physical laws and biological processes, we might define this as free will. However, if ultimate causes lie beyond our universe, our "choices" could be mere epiphenomena, a subjective interpretation of an inevitable unfolding, raising profound questions about agency and autonomy.

Consider the implications for consciousness. If determinism holds sway and our universe is a predictable outcome of external cosmic processes, does this diminish the significance of our conscious experience? Is our awareness, introspection, and search for meaning merely an elaborate biological mechanism following a predetermined pattern? Or could consciousness itself introduce a unique form of causality, operating within the broader cosmic framework yet allowing for a degree of self-determination beyond simple deterministic chains? Some interpretations of quantum mechanics, particularly concerning the observer's role, suggest consciousness might actively shape reality, albeit within cosmic constraints, rather than passively receiving cosmic causality.

The black hole universe hypothesis forces us to confront the possibility that our perception of time and causality is inherently limited by our position within this cosmic structure. We experience time linearly, with sequential events. However, within the extreme spacetime curvature near a black hole, time dilation occurs, and the concept of a universal "now" breaks down. If our universe originated from such a region, our linear experience of time might be a feature of our internal cosmic environment, not a universal constant. This could imply that the deterministic chain of events, from our perspective, might appear differently from an external viewpoint, or that past and future events are in some way already determined or interconnected in ways we cannot comprehend.

The scientific challenge lies in bridging the gap between our internal, observable universe and the hypothetical external reality governing its existence. If our universe is a black hole's interior, understanding the causal links to the parent universe necessitates a deep grasp of quantum gravity. Theories unifying general relativity and quantum mechanics, like string theory or loop quantum gravity, are essential for probing these extreme conditions. Within these theoretical frameworks, we might discover whether cosmic causality is absolute or if probabilistic or even agentic elements allow for genuine freedom within the grand cosmic machinery.

Framed within a black hole universe, the question of determinism versus free will becomes a question about the ultimate nature of reality itself. Is reality a vast, predictable clockwork mechanism, or a dynamic, emergent process imbued with chance and perhaps an intrinsic drive towards complexity and consciousness? The black hole hypothesis doesn't offer an immediate answer but profoundly enriches the debate by expanding the scope of potential causal influences beyond our current observational reach. It compels us to consider that the origins of our existence, and thus the fundamental nature of causality and our capacity for free action, might be rooted in the very fabric of a larger, unseen cosmos. This is a frontier where physics and philosophy converge, urging us to look beyond our observable universe to understand the deepest questions about our place and agency within the grand cosmic narrative. The intricate causal web defining our reality may, in fact, extend infinitely, with each thread woven from the profound and mysterious processes that birth and sustain universes.

The idea that our universe might exist within a black hole in a larger, external universe profoundly alters our search for meaning. For ages, humanity has sought purpose by observing the cosmos, developing complex cosmologies, and exploring consciousness. The black hole universe hypothesis introduces a new dimension, shifting our perspective from isolated observation to one of intricate, nested belonging.

This concept inspires immense awe and wonder. If our universe is inside a black hole, then our reality—spacetime, matter, energy, and fundamental laws—are all products of extreme gravitational physics in a parent cosmos. The Big Bang, in this context, might not be an absolute beginning but an emergent phenomenon, a transition from a singularity or a dense region of spacetime. This powerful origin, shaped by forces beyond our current understanding, gives our universe a mythical genesis. The sheer scale of a universe within a universe humbles us, reminding us that our perceived immensity is a speck in a grander cosmic tapestry. This humility can inspire appreciation for the unique conditions that allowed our existence.

Furthermore, this hypothesis fosters a deep sense of interconnectedness. If our universe results from a parent black hole's dynamics, our existence is inextricably linked to that external reality. We are not just observers but participants in a larger system, suggesting a cosmic solidarity where the conditions for our universe are tied to the host universe's physical laws and evolution. This interconnectedness highlights the delicate balance of physical constants enabling stars, galaxies, and life. It implies our universe isn't random but a specific outcome of a larger cosmic process, making our existence a product of this grand, interconnected causal chain. This realization can inspire a sense of responsibility, not just for our actions within our universe but for our universe's abstract role within the larger cosmic structure, even if unperceivable.

The search for meaning, often driven by the desire for a unique human purpose, might evolve if our universe is a black hole's interior. Our quest would shift from seeking an external, divine decree to understanding our role as emergent phenomena within an alien cosmic environment. Our purpose might not be predetermined but rather the unfolding of complexity and consciousness within our reality's specific parameters. Consciousness itself—self-awareness and the pursuit of understanding our origins—could be a significant contribution to the larger cosmic narrative, an instance of the universe becoming aware through us. This internal drive for understanding becomes a testament to existence's potential, regardless of its ultimate cosmic housing, shifting focus from external validation to the inherent value of experience, consciousness, and the human pursuit of knowledge.

This perspective also challenges anthropocentric views of meaning. While many traditions have placed humanity at the center of creation, the black hole universe hypothesis displaces us from such a privileged position. We are not the cosmos's sole focus but a remarkable, perhaps improbable, emergent feature of a vaster reality. This can be a humbling yet liberating realization, freeing us from a singular, imposed purpose. It allows us to define our own meaning through our actions, relationships, and contributions to life's flourishing and understanding within our cosmic sphere. The meaning we create is not diminished but grounded in our lived experience and our capacity for love, compassion, and intellectual pursuit.

Consider the implications for life and consciousness. If our universe emerged from a black hole's highly energetic and complex environment, it suggests that life's necessary conditions might be intrinsically linked to such extreme astrophysical phenomena. This could mean life's emergence is not an anomaly but a natural, albeit complex, consequence of universal processes. Our consciousness, then, could be a sophisticated product of these processes, an advanced stage in the universe's journey toward self-understanding. The capacity for subjective experience, emotion, creativity, and moral reasoning enriches existence within our universe, adding layers of meaning significant in their own right, regardless of the larger cosmic architecture. The quest for meaning thus intertwines with cultivating these uniquely human qualities and celebrating the emergent complexity that life represents.

The concept also encourages a re-evaluation of finitude and impermanence. If our universe is within a black hole, its existence is likely bounded. However, this potential impermanence does not negate the value or meaning of our existence within our temporal and spatial limits. Instead, it can imbue our present moments with greater urgency and significance. Knowing that our cosmic home has a beginning and potentially an end—or a transformation—encourages us to live fully, cherish experiences, and strive to leave a positive legacy. The pursuit of knowledge, creation of art, community building, and fostering empathy all become vital endeavors contributing to the meaning of our transient cosmic epoch.

Moreover, the black hole universe notion offers a novel perspective on transcendence. Traditionally, transcendence often involves reaching beyond the material world to a spiritual or divine realm. In this context, transcendence could be interpreted as understanding and appreciating the larger cosmic reality that contains us. It is a form of intellectual and philosophical transcendence, achieved through scientific inquiry and contemplation, allowing us to grasp our place within a hierarchy of cosmic structures. The search for meaning becomes, in part, a journey towards this expanded awareness, a process of shedding our provincial cosmic perspective and embracing a more inclusive understanding of reality.

The hypothesis of our universe existing within a black hole suggests a form of cosmic determinism, yet meaning can still be found within this framework. If the universe operates on predetermined laws, then our existence and quest for meaning are inherent parts of this grand design. Meaning, therefore, arises from our participation in this process—our ability to understand, experience, and contribute to cosmic evolution. It's not about absolute freedom from causality, but rather the intricate causal web that fosters consciousness and the pursuit of knowledge.

This concept profoundly shifts our philosophical perspective, moving us from an isolated existence in a vast void to a deeply integrated reality within a dynamic cosmic structure. Such a reorientation can inspire humility, awe, and a sense of interconnectedness, reshaping our understanding of purpose. Rather than diminishing the human drive for significance, this cosmic view amplifies it, challenging and enriching us with the realization that our universe might be far more complex and interconnected than we imagined. The search for meaning transcends internal exploration, becoming a cosmic journey into a mystery beyond our observable horizon. Embracing this grander context may provide the most profound answers to fundamental human questions about our identity and purpose.

The bold proposition that our universe is a black hole within a larger cosmos is more than a cosmological conjecture; it's a powerful testament to enduring scientific curiosity. This innate human drive—the relentless urge to question, explore, and expand our knowledge—is the very engine of scientific progress. It mirrors the ancient impulse that led early humans to gaze at the stars, sparking myths and, eventually, sophisticated tools and theories to probe extreme cosmic environments. The black hole universe hypothesis, with its mind-bending implications, represents a modern expression of this ancient drive. It compels us to confront the limits of our current understanding, not as points of despair, but as frontiers for exploration, inviting innovation and a reconceptualization of reality itself.

Throughout scientific history, paradigm shifts, such as the Copernican revolution or the development of quantum mechanics, have faced initial resistance, challenging deeply held assumptions. The heliocentric model, for instance, took decades of observation and argument to displace geocentric thought. Similarly, quantum theory, with its counter-intuitive principles, has taken generations to be partially assimilated. The black hole universe hypothesis, though perhaps more speculative, aligns with this lineage of bold, paradigm-shifting ideas. It demands intellectual flexibility and a willingness to consider initially outlandish concepts, as these explorations often reveal deeper truths about the universe.

Formulating and investigating such hypotheses fundamentally involves grappling with the limits of our knowledge. Our current scientific framework, powerful as it is, is based on observations and theories from within our observable universe. When we contemplate phenomena beyond this empirical domain—such as what lies beyond an event horizon or the conditions before the Big Bang—we encounter conceptual hurdles. These aren't necessarily barriers but rather indicators of where our models might be incomplete or require radical extensions. The black hole universe idea pushes us to confront these edges, prompting questions about the physics governing the "parent" universe and the fundamental constraints of such a reality.

Even without definitive answers, the act of asking these questions is intrinsically valuable. Striving to comprehend the incomprehensible often refines our methods and deepens our understanding of existing concepts. For example, theoretical work on black holes, including the event horizon and information paradox, has expanded general relativity and quantum field theory, offering new insights into gravity, spacetime, and information. The black hole universe hypothesis extends this intellectual exploration, prompting us to consider how black hole properties might relate to the origin and structure of entire universes. This cross-pollination of ideas, where concepts from one field inform new avenues of inquiry, is a hallmark of robust scientific exploration.

Furthermore, scientific endeavor is committed to falsifiability—the principle that a theory must be capable of being proven wrong. While the black hole universe hypothesis presents challenges to direct empirical verification, it also stimulates the development of indirect observational strategies and theoretical constraints. Astrophysicists might search for subtle gravitational imprints in cosmic microwave background radiation or specific spectral signatures in distant galaxies, hinting at nested cosmic structures. Theoretical physicists might develop mathematical models incorporating the possibility of a black hole universe, exploring their consistency with established physical laws. This constant interplay between theory and observation, speculative thought and rigorous testing, keeps science dynamic and grounded.

The pursuit of understanding the universe, especially when confronting profound ideas like a black hole universe, also highlights the limitations of our current conceptual tools. Our language, intuitions, and mathematical frameworks are largely shaped by our experiences in everyday, low-gravity reality. Concepts like singularity, event horizon, and infinite density push the boundaries of our intuitive grasp. We rely on abstract mathematical constructs and analogies, but a fundamental gap remains between our everyday understanding and the reality they describe. This gap isn't a failure but an indication of the vast unknown and the potential for future conceptual revolutions. Perhaps a deeper understanding of consciousness or entirely new mathematical formalisms will be necessary to fully grasp the implications of such cosmological scenarios.

Scientific curiosity naturally pushes us to explore beyond what we can currently observe. Our observable universe, defined by the reach of light since the Big Bang, might be a small part of a much larger, even infinite, reality. The black hole universe hypothesis suggests our universe could be contained within a larger structure, prompting epistemic humility. This view implies our current scientific understanding is like a map of a single continent, unaware of the rest of the world.

This humility isn't a deterrent but a powerful motivator, reminding us that the universe is likely far more complex than we imagine. Each unanswered question and paradox encourages deeper investigation. The scientific journey is a continuous process of refinement and discovery, not a search for a final answer. The black hole universe hypothesis serves as a waypoint, a thought experiment that encourages us to view our cosmic home from new perspectives. It prompts questions: what are the properties of such a black hole, what is the nature of the external universe it inhabits, and what are the potential observational consequences?

The limitations of knowledge also extend to philosophy. If our universe is nested within another, the questions of ultimate origins and purposes gain new dimensions. While this might seem to diminish humanity's central role, it can also be liberating, allowing us to focus on creating meaning within our own existence and appreciating our consciousness and capacity for understanding, regardless of the ultimate cosmic context. The human drive for meaning is an emergent property of complex systems, a testament to the universe's capacity for self-awareness through conscious beings.

The scientific community thrives on challenging established ideas. The black hole universe hypothesis exemplifies this spirit, stemming from rigorous theoretical physics and representing an extension of established science, not a defiance. It reflects the natural progression of scientific inquiry where new insights and data lead to new questions and further research. The fact that such a hypothesis is seriously considered highlights the intellectual dynamism and openness of modern physics.

Ultimately, the pursuit of these profound questions defines the human experience of exploring the cosmos. We are inherently driven to understand. Even when faced with the limits of our knowledge, the urge to probe and imagine persists. The black hole universe hypothesis, though speculative, powerfully reminds us that our quest for knowledge is an ongoing, dynamic process. It celebrates scientific curiosity, acknowledging that the most exciting discoveries often lie at the edge of our understanding, inviting us to expand our horizons and embrace the universe's profound mysteries. This relentless pursuit of answers is not just a scientific endeavor but a fundamental aspect of being human.

CHAPTER 10

The Future of Cosmology: New Horizons

The concept of our universe existing within a black hole in a larger cosmos is more than just a theoretical idea; it exemplifies the powerful nature of scientific curiosity. This fundamental human drive—to question, explore, and expand our knowledge—is the very force behind scientific advancement. It mirrors the ancient human wonder at the night sky, which spurred the creation of myths and, eventually, the sophisticated tools and theories we now use to examine the universe's most extreme conditions. The black hole universe hypothesis, with its complex implications, is a modern expression of this age-old impulse. It challenges the limits of our current understanding, presenting them not as obstacles, but as opportunities for exploration, innovation, and a redefinition of reality.

Consider the historical progression of scientific discovery. Every major shift, from the Copernican revolution to the advent of quantum mechanics, initially faced resistance and challenged deep-seated beliefs about our place in the cosmos. The heliocentric model, which removed Earth from the universe's center, was not immediately embraced; it took decades of observation, calculation, and persuasive arguments to overcome the ingrained geocentric view. Similarly, quantum theory, with its inherent strangeness, has taken generations to be even partially understood. The black hole universe hypothesis, while perhaps even more speculative, follows in this tradition of bold, transformative ideas. It requires similar intellectual flexibility and a willingness to consider initially outlandish concepts, as these explorations often reveal deeper truths about the universe.

The development and investigation of such hypotheses are essentially about confronting the boundaries of our knowledge. Our current scientific framework, while incredibly robust, is based on observations and theories derived from within our observable universe. When we consider phenomena outside this empirical realm—such as the nature of the region beyond an event horizon or the conditions before the Big Bang—we inevitably encounter conceptual difficulties. These are not necessarily impediments to understanding, but rather indications of where our current models may be incomplete or require radical expansion. The black hole universe idea compels us to confront these boundaries, to ask what might lie beyond the event horizon, not just in terms of physics, but in terms of conceptual frameworks. What kind of physics would govern the "parent" universe containing our black hole universe? What are the fundamental constraints on such a reality?

The very act of posing these questions, even without immediate definitive answers, holds intrinsic value. It is in the pursuit of comprehending the incomprehensible that we often refine our methods and deepen our understanding of existing concepts. For example, theoretical work surrounding black holes, including the event horizon and the information paradox, has pushed the limits of general relativity and quantum field theory, leading to new insights into gravity, spacetime, and information. The black hole universe hypothesis extends this intellectual journey, prompting us to consider how black hole properties might relate to the origin and structure of entire universes. This interdisciplinary exchange of ideas, where concepts from one area of physics inform new lines of inquiry, is a hallmark of vigorous scientific exploration.

Furthermore, scientific endeavor is characterized by a commitment to falsifiability, meaning a scientific theory must be capable of being disproven. While the black hole universe hypothesis inherently presents challenges to direct empirical verification, it also encourages the development of indirect observational strategies and theoretical constraints. Astrophysicists might search for subtle gravitational imprints in the cosmic microwave background radiation, or specific spectral signatures in distant galaxies that could suggest the presence of such nested cosmic structures. Meanwhile, theoretical physicists might work to develop mathematical models that incorporate the possibility of a black hole universe, exploring the consistency of such models with other established physical laws. This constant interaction between theory and observation, between speculative thought and rigorous testing, keeps science dynamic and grounded.

The quest to comprehend the universe, especially profound concepts like a black hole universe, underscores the limitations of our current intellectual tools. Our understanding, language, and mathematical frameworks are rooted in our everyday, low-gravity experiences. Concepts such as singularity and event horizon challenge our intuitive grasp, requiring abstract mathematical constructs and analogies. This gap signifies not a failure, but the immense unknown and the potential for future conceptual breakthroughs, possibly through a deeper understanding of consciousness or new mathematical formalisms.

Scientific curiosity propels us beyond our observable horizons. The observable universe, defined by the reach of light since the Big Bang, might be a mere fraction of a much larger, even infinite, cosmos. The black hole universe hypothesis posits that our observable universe could reside within a structure that is part of a grander, potentially unknowable, reality. This introduces a sense of epistemic humility, suggesting our current scientific knowledge is like a map of a single continent, unaware of the rest of the planet.

This humility, far from deterring the pursuit of knowledge, serves as a powerful motivator. It reminds us of the universe's likely strangeness and complexity, making each unanswered question an invitation to deeper investigation. The scientific journey is a continuous process of refinement, discovery, and re-evaluation, not a search for a final understanding. The black hole universe hypothesis is a stimulating thought experiment that prompts us to consider new perspectives on our cosmic home. It leads us to ask about the properties of such a black hole, the nature of the external universe it inhabits, and its potential observational consequences.

The limitations of knowledge also extend to philosophy. If our universe is nested within another, questions of ultimate origins and purposes take on a new dimension. This perspective, while seemingly demoting humanity, can also be liberating, allowing us to focus on creating meaning within our existence and appreciating our consciousness and capacity for understanding, irrespective of the ultimate cosmic context. The human drive for meaning is an emergent property of complex systems, a testament to the universe's capacity for self-awareness.

The scientific community thrives on challenging established ideas. The black hole universe hypothesis, a product of rigorous theoretical physics and decades of research, embodies this spirit. Its exploration is an extension of established science, representing the natural progression of inquiry where new insights drive further research. The very consideration of such a hypothesis demonstrates the intellectual dynamism of modern physics.

Ultimately, exploring profound and potentially unanswerable questions defines the human experience of understanding the cosmos. We are inherently driven to understand, and even when faced with the limits of our knowledge, the urge to probe and imagine persists. The black hole universe hypothesis, though speculative, powerfully reminds us that our quest for knowledge is an ongoing, dynamic process. It celebrates scientific curiosity, acknowledging that the most exciting discoveries often lie at the edge of our understanding, inviting us to expand our horizons and embrace the universe's mysteries. This relentless pursuit of answers is not just a scientific endeavor but a fundamental aspect of being human.

Testing the Black Hole Universe Hypothesis: Future Observational and Theoretical Frontiers

Despite its abstract theoretical roots, the black hole universe hypothesis is a scientific endeavor that demands rigorous testing through observable consequences and established physical laws. While directly verifying our universe's placement within a black hole poses significant challenges, the path forward involves identifying subtle, indirect signatures and developing theoretical frameworks to support or refute this radical cosmology. The future of cosmology, in this regard, necessitates innovative observational strategies and continued theoretical advancements to assess this intriguing idea's viability.

One promising avenue for testing the black hole universe hypothesis is gravitational wave astronomy. With increasingly sensitive detectors like LIGO, Virgo, KAGRA, and future observatories like LISA, we will gain unprecedented access to gravitational wave events. The hypothesis suggests our universe is the interior of a black hole situated within a larger, external cosmos. The dynamics of this "parent" universe, and the nature of the black hole containing us, could leave detectable gravitational wave imprints. For instance, violent events in the parent universe, such as mergers of massive objects or phase transitions, could propagate as gravitational waves. Detecting such waves, particularly those with characteristics significantly different from typical astrophysical sources within our universe, could offer a crucial clue.

More specifically, the black hole's event horizon and core singularity might manifest uniquely. Classical general relativity predicts an inescapable event horizon and an infinite spacetime curvature singularity. However, quantum gravity effects are expected to modify these descriptions. If our universe is inside a black hole, the quantum nature of its "horizon" and the fate of matter that "formed" it could be imprinted on the gravitational wave background. A subtle, persistent hum of gravitational waves – a stochastic background – with specific spectral properties not easily explained by known astrophysical sources could be a "smoking gun." The frequency and amplitude distribution of such a background might reveal information about the mass, spin, and internal structure of the black hole hosting our universe. Furthermore, the process by which our universe "formed" within this black hole might have generated specific gravitational wave signals. If, for example, our universe emerged from a highly energetic event in the parent universe that subsequently collapsed into a black hole, the gravitational wave signature of that initial event might be preserved in the background.

Another crucial area for testing the black hole universe hypothesis involves precision measurements of the Cosmic Microwave Background (CMB) radiation. The CMB, the afterglow of the Big Bang, is a remarkably uniform snapshot of the early universe, yet it contains tiny temperature fluctuations and polarization patterns that seeded all cosmic structures. The black hole universe model could potentially influence these primordial fluctuations in several ways.

The structure of a black hole's event horizon could provide clues to whether our universe exists within one. Instead of a perfectly smooth sphere, the boundary might have anisotropies or patterns due to its formation or quantum effects. For example, if our universe originated from matter falling into a black hole, the distribution of that matter could have left non-Gaussianities or specific correlations in the Cosmic Microwave Background (CMB) temperature fluctuations. Theoretical models suggest that a black hole event horizon acting as our universe's boundary might possess unique properties reflected in CMB polarization. Detecting subtle hemispheric asymmetry or unusual correlations in CMB polarization that deviate from standard inflationary cosmology would be significant evidence.

This hypothesis could also explain anomalies in the CMB, such as the "cold spot" or the alignment of large-scale structures, which are often attributed to statistical fluctuations or measurement errors. If our universe is inside a black hole, the internal matter and energy distribution might create preferred directions or regions of varying density, leading to these CMB anomalies. The cosmic web, the universe's large-scale filamentary structure, could also be influenced by the parent universe's gravitational pull or the black hole's internal shape.

Beyond the CMB, large-scale structure surveys offer another test. The distribution of galaxies and galaxy clusters, forming the cosmic web, is a result of initial density fluctuations amplified by gravity. If our universe is within a black hole, the gravitational environment might subtly differ from standard cosmology, potentially affecting the universe's expansion history. The way matter clumps on large scales might reveal discrepancies, such as subtle deviations in galaxy correlation functions or the growth rate of cosmic structures, requiring precise measurements of vast numbers of galaxies.

The black hole universe hypothesis might also offer new insights into the cosmological constant problem and dark energy. Standard cosmology attributes the universe's accelerated expansion to dark energy, a mysterious force with negative pressure, whose incredibly small observed magnitude compared to theoretical predictions is a long-standing puzzle. If our universe is inside a black hole, its expansion could be driven by the black hole's internal dynamics or by effects from an external universe. The cosmological constant might then be a residual effect of the parent universe's properties or the black hole's quantum state. Future observations measuring dark energy's equation of state and its evolution could reveal deviations from standard models, supporting this alternative view.

This hypothesis necessitates a deeper understanding of quantum gravity. General relativity breaks down at a black hole's singularity, requiring a quantum theory of gravity, much like understanding the early universe and the Big Bang singularity. If our universe is indeed the interior of a black hole, a complete theory of quantum gravity might naturally lead to such a scenario. Theoretical physicists are actively pursuing various approaches to quantum gravity, like string theory and loop quantum gravity. A consistent and predictive quantum gravity theory could provide a framework for the black hole universe hypothesis. For example, some theories suggest that quantum effects might "resolve" singularities, leading to a smooth, albeit exotic, internal structure for black holes. If such a theory predicts a black hole interior capable of containing a universe similar to our own, it would significantly support the hypothesis.

Furthermore, the information paradox associated with black holes, concerning the fate of information that falls in, could be indirectly relevant. If our universe is within a black hole, the fate of information crossing its apparent boundary (the parent black hole's event horizon) becomes critical. Proposed resolutions to the information paradox, like the holographic principle, suggest information might be encoded on the event horizon. If our universe is "inside" such a horizon, the information constituting our reality would need to be consistent with that boundary's properties, potentially constraining the physical laws and structures we observe.

Falsification is crucial for the black hole universe hypothesis to be a viable scientific theory; it must make testable predictions that could be proven wrong. While direct observation of our universe's "outside" is impossible, indirect consequences are key. These might include:

- **Deviations from the standard cosmological model:** Any significant, otherwise unexplainable departure from the Lambda-CDM model's predictions in CMB data, large-scale structure, or expansion history would be strong evidence.
- **Specific gravitational wave signatures:** The detection of a unique stochastic gravitational wave background or peculiar features in individual gravitational

wave events characteristic of a black hole interior or its formation.

- **Anomalous CMB patterns:** Unexplained patterns in the CMB, especially in polarization, predicted by black hole universe models.
- **Properties of dark energy and cosmic acceleration:** If dark energy's nature fundamentally differs from a cosmological constant or a simple scalar field, it could suggest an external influence or a different cosmological mechanism.

Differentiating between a universe inside a black hole and one merely mimicking some of its properties (e.g., accelerated expansion) will be a significant challenge. Precisely measuring cosmological parameters and searching for subtle deviations will be critical. Future generations of telescopes and gravitational wave detectors with enhanced sensitivity and resolution will be indispensable.

The theoretical work is equally demanding. Cosmologists and theoretical physicists must develop detailed, self-consistent models of how a universe could form and evolve within a black hole, understanding the singularity, the event horizon from a quantum perspective, and the dynamics of matter and energy in such an environment. It also requires exploring implications for fundamental physics, such as spacetime's nature, fundamental forces, and the existence of other dimensions or realities.

In summary, testing the black hole universe hypothesis requires a multi-pronged approach, integrating cutting-edge observational astronomy with advanced theoretical physics. Gravitational wave astronomy, precision CMB measurements, large-scale structure surveys, and a deeper understanding of dark energy and quantum gravity are all critical. While speculative, pursuing its verification or refutation promises to push our understanding, potentially leading to revolutionary insights into reality, our cosmos's origin, and our place in the cosmic tapestry. It embodies the scientific spirit: questioning the impossible, seeking evidence in unexpected places, and continually refining our universe models, regardless of the profound implications.

Our universe may exist within a black hole, a theory that relies on increasingly precise observations and measurements. While past instrumentation limited such ideas to speculation, advancements in next-generation telescopes and detectors are poised to revolutionize observational cosmology. These technological leaps are not mere improvements; they represent a paradigm shift in our ability to probe the universe's subtle features and, critically, to validate or refute hypotheses like the black hole universe.

A central component of this observational revolution is the upcoming generation of space-based observatories. Free from atmospheric distortion, these platforms offer unparalleled clarity and sensitivity. Successors to the James Webb Space Telescope (JWST), for example, will boast even greater sensitivity, wider spectral coverage, and enhanced angular resolution, enabling unprecedented fidelity in observing the Cosmic Microwave Background (CMB). The black hole universe hypothesis could manifest in subtle CMB anisotropies or polarization patterns—features only hinted at by current telescopes. Future instruments, by meticulously mapping these primordial fluctuations, might detect patterns inconsistent with standard inflationary cosmology, such as specific correlations or deviations from Gaussianity, suggesting the influence of a larger cosmic structure or the unique geometry of a black hole's interior.

Furthermore, the search for specific CMB signatures may involve examining subtle spectral distortions or anomalies in polarization distribution. Theories positing a black hole universe often include unique boundary conditions or the influence of a parent universe, which could leave imprints on the CMB currently indistinguishable from noise. Next-generation telescopes, equipped with highly sensitive bolometers and sophisticated signal processing, will be able to extract these faint signals from overwhelming noise. This includes identifying gravitational lensing effects on the CMB not attributable to known structures, or polarization patterns indicating a preferred directionality or asymmetry originating from our cosmic confinement's "boundary," if it behaves like an event horizon.

Beyond the CMB, future space telescopes will refine measurements of large-scale structure and the universe's expansion history. The cosmic web, a vast network of galaxies and dark matter, arises from initial density fluctuations amplified by gravity. Upcoming observatories will significantly enhance the precise mapping of this structure, including galaxy distribution and dark matter halo nature. These surveys will allow cosmologists to test predictions about how matter clumps and evolves within an environment potentially shaped by a black hole's gravitational influence. For instance, if our universe is a black hole's interior, matter distribution might subtly deviate from the standard Lambda-CDM model on the largest scales. Future galaxy surveys, utilizing techniques like redshift-space distortions and weak gravitational lensing, will provide far more accurate measurements of these large-scale correlations, offering a sensitive probe of underlying cosmological geometry.

The ability of these telescopes to probe faint light from the most distant objects is also paramount. By observing the spectral characteristics of the earliest stars and galaxies, we can glean information about the universe's initial conditions and evolution. If our universe is a black hole's interior, its formation might have involved unique nucleosynthesis pathways or a different distribution of primordial elements. Future telescopes could detect subtle variations in light element abundance or the presence of exotic particles not accounted for in standard Big Bang nucleosynthesis, potentially offering clues to our cosmic origins within a different framework.

The revolution in gravitational wave astronomy presents perhaps the most direct and powerful new avenue for testing the black hole universe hypothesis. While LIGO, Virgo, and KAGRA have provided groundbreaking detections of black hole and neutron star mergers, their sensitivity is limited to specific frequency ranges. The next generation of gravitational wave detectors, both ground-based and space-based, promises to open entirely new windows into the gravitational universe.

The Laser Interferometer Space Antenna (LISA), a space-based observatory designed to detect gravitational waves in the millihertz frequency range, is particularly relevant. This is precisely where gravitational waves from supermassive black hole mergers, extreme mass ratio inspirals (EMRIs), and potentially the relic gravitational wave background from the very early universe are expected. If our universe is a black hole's interior, its formation or evolution could have generated gravitational waves within LISA's sensitive band. For example, if the parent universe underwent energetic events, or if our universe's formation involved matter inspiraling into a pre-existing black hole, LISA might detect a continuous gravitational wave hum or distinct merger events not attributable to known astrophysical phenomena within our universe.

The nature of the event horizon itself, a defining feature of a black hole, could also be probed through gravitational waves. While classical general relativity describes a smooth horizon, quantum gravity is expected to modify this, perhaps introducing a "fuzzy" horizon or a region of highly distorted spacetime. If our universe is such an interior, its "boundary" might have specific properties imprinted on gravitational waves. Detecting gravitational waves with subtle deviations from expected waveforms produced by classical black hole mergers—perhaps related to echoes or modified ringdown phases—could be a signature of a quantum horizon or an exotic interior. The sensitivity of future detectors like the Einstein Telescope and Cosmic Explorer on the ground, and LISA in space, will be crucial in discerning these subtle effects.

The search for a stochastic gravitational wave background, a persistent hum from unresolved sources or the early universe, is critically important. If our universe originated from a cosmological-scale black hole or exists within a larger structure that experienced such events, a unique gravitational wave background could be present. Its spectral shape and amplitude would offer direct insights into the physics of the parent universe or the nature of the black hole containing ours. Distinguishing this background from astrophysical sources in our universe will demand highly precise measurements and advanced data analysis techniques, which are currently under development for next-generation instruments.

These advancements could also shed light on the nature of dark energy and cosmic acceleration. If our universe's accelerated expansion results from its embedding within a larger structure or a black hole, rather than an intrinsic dark energy field, then future observations precisely measuring dark energy's equation of state and evolution might reveal deviations from standard cosmological models. Observing an acceleration that deviates from a cosmological constant or exhibits unusual behavior on vast scales could indirectly support an external influence or a different cosmological paradigm. Next-generation telescopes capable of high-precision measurements of supernovae, baryon acoustic oscillations, and weak lensing will be crucial for this endeavor.

Our understanding of the universe is inherently tied to technological progress. Each advancement in telescope design, detector sensitivity, and data analysis opens new observational windows, enabling us to test previously untestable hypotheses. Next-generation instruments promise the precision needed to scrutinize the cosmos's most subtle features, from the faint echoes of the Big Bang to the gravitational whispers of cosmic mergers. These technological marvels are not merely tools; they are extensions of our curiosity, allowing us to explore existence's deepest mysteries.

Unraveling the universe's mysteries, especially the audacious hypothesis that our cosmos might reside within a black hole, hinges on advancing our understanding of quantum gravity. This theoretical discipline aims to reconcile general relativity, which governs gravity and large-scale spacetime, with quantum mechanics, which describes matter and energy at the smallest scales. The transition from the singularity predicted at the heart of a classical black hole to our universe, with its vast structures and complex physics, necessitates a quantum gravitational description. Without it, the concept of a universe born from or existing within such an extreme gravitational environment remains speculative.

Theoretical physicists are actively developing various quantum gravity approaches, each offering a unique perspective on the fundamental nature of reality. String theory, a well-developed candidate, posits that the universe's fundamental constituents are tiny, vibrating strings. Different vibration modes correspond to different fundamental particles and forces, including gravity. Within this framework, black holes can be understood as complex configurations of these fundamental strings, and their interiors, along with processes leading to universe creation, are subjects of intense investigation. The "landscape" of possible vacuum states in string theory suggests an enormous number of potential universes. Understanding which, if any, could describe our own, or the one that birthed ours, is a formidable challenge. However, if string theory accurately describes quantum gravity, it could provide a roadmap for how a universe might emerge from the extreme conditions associated with a black hole, potentially resolving the singularity problem and offering a framework for calculating observable consequences.

Loop Quantum Gravity (LQG) offers an alternative, quantizing spacetime itself as a granular, atomic structure of discrete "loops." Spacetime, in this view, is a quantum weave, not a smooth continuum. Applying LQG to a black hole's interior or the Big Bang's initial moments (analogous to formation within a parent black hole) suggests the singularity is resolved by a "quantum bounce." This bounce, a reversal of gravitational collapse driven by quantum effects, could have initiated our universe. If LQG is correct, echoes of this bounce or the specific properties of the quantized spacetime characterizing our universe's genesis might be imprinted on cosmological observables. The challenge for LQG proponents is to derive testable predictions from these abstract quantum geometric considerations for astronomical verification.

Other promising avenues include Causal Dynamical Triangulations (CDT), which uses a discrete approximation of spacetime to explore quantum gravitational effects, and asymptotically safe gravity, which proposes that gravity remains well-behaved at high energies due to a non-trivial ultraviolet fixed point in its renormalization group flow. Each framework attempts to tackle what happens at energy densities and spacetime curvatures far exceeding current experimental access—conditions expected at a black hole's heart or the universe's beginning.

The crucial interplay between theoretical developments and observational cosmology will ultimately drive progress. Elegant mathematical theories must make verifiable predictions. For the black hole universe model to move from speculation to established science, quantum gravity theories must provide a mechanism for its formation and evolution, and these mechanisms must lead to observable signatures. For example, if our universe emerged from a bounce within a parent black hole's interior, the bounce's specific properties—its energy scale, duration, and generated quantum fluctuations—could have left tell-tale signs in the Cosmic Microwave Background (CMB). Subtle deviations from standard inflationary predictions, such as specific non-Gaussianity patterns in the CMB or a particular spectrum of primordial gravitational waves, could be the smoking gun.

Consider the detailed predictions that might emerge from a quantized spacetime. In loop quantum cosmology, a specific prediction is the generation of a stochastic gravitational wave background during the Big Bounce phase. The frequency spectrum and amplitude of this background are sensitive to the quantum gravity theory's details and the bounce mechanism. Future gravitational wave observatories, such as LISA (Laser Interferometer Space Antenna) and advanced ground-based detectors like the Einstein Telescope, are designed to probe frequency ranges that could potentially detect such a background. If the black hole universe hypothesis is correct and our universe originated from such a bounce, the gravitational wave signature would directly confirm the quantum gravitational processes involved in its birth. Furthermore, the theory might predict specific deviations in the matter power spectrum or the large-scale structure distribution that differ from standard inflationary models, providing another avenue for observational verification.

The challenge lies in the fact that quantum gravity effects are typically expected to manifest at the Planck scale—incredibly small distances and unimaginably high energies. However, theories like cosmic inflation or a Big Bounce can amplify these microscopic effects to macroscopic scales, making them potentially observable in the early universe. The black hole universe model, by proposing a specific origin and embedding for our cosmos, provides a concrete context for these quantum gravitational considerations. If our universe is indeed a black hole's interior, the very boundary conditions of this cosmic abode, dictated by quantum gravity, could influence its evolution in ways not captured by standard cosmological models. For instance, the "horizon" analogous to a black hole's event horizon might be a region of quantum foam or a transition zone with specific thermodynamic properties, potentially affecting cosmic expansion's late-time behavior or dark energy's properties.

The refinement of quantum gravity theories is a painstaking process involving mathematical rigor and constant dialogue with observational data. Astronomers are pushing observational boundaries, seeking increasingly precise measurements of the CMB, galaxy distribution, and the universe's expansion history. Theoretical cosmologists, armed with evolving quantum gravity tools, then attempt to interpret these observations within their theoretical frameworks. If data deviates from standard cosmological models' predictions, it opens the door for alternative explanations, including those rooted in quantum gravity and unconventional cosmological scenarios like the black hole universe.

Observational Tests for a Black Hole Universe and the Role of Quantum Gravity

Future, highly precise measurements of the Cosmic Microwave Background (CMB) could provide crucial evidence for or against the black hole universe scenario. If these measurements reveal correlations or anisotropies that standard inflationary cosmology cannot explain, theorists will investigate whether specific quantum gravity models, when applied to the black hole universe, can reproduce these patterns. This includes exploring how the curvature of a higher-dimensional spacetime containing the black hole, or the quantum nature of its event horizon analogue, might influence primordial fluctuations. Similarly, discrepancies in cosmological parameters, like the Hubble constant or the amplitude of matter fluctuations, could signal a departure from the standard Lambda-CDM model, necessitating a re-evaluation that considers a quantum gravitational origin for our universe.

The direct detection of primordial gravitational waves would be revolutionary, offering direct insights into the physics of the very early universe. If our universe originated from a Big Bounce within a black hole, the quantum gravitational processes involved would uniquely shape the spectrum of these waves, potentially differentiating it from predictions of standard inflation. Theorists are actively calculating these specific spectral signatures within various quantum gravity frameworks, such as Loop Quantum Gravity (LQG) and string theory, in the context of the black hole universe. Comparing these theoretical predictions with future observational data will be a critical test.

Quantum Gravity: Essential for Understanding Cosmic Origins

Developing a successful theory of quantum gravity is not merely an academic exercise; it is fundamental for any robust model of cosmic origins, especially unconventional ones like the black hole universe hypothesis. It is essential for understanding how spacetime and matter behave under the extreme conditions present at the universe's beginning, or at the "boundary" of our cosmic existence if it resides within a larger structure. As our observational capabilities advance, they impose increasingly strict constraints on theoretical models. The quantum gravity theory that can successfully explain the observed universe, including any anomalies deviating from standard cosmology, and provide a consistent framework for a black hole-originated cosmos, will be the one to prevail. The synergy between theoretical advancements in quantum gravity and the relentless pursuit of observational precision is the driving force behind our quest to understand the universe's deepest secrets. The black hole universe model, though radical, strongly motivates pushing the boundaries of both theoretical and observational cosmology, demanding a quantum description of gravity that can bridge the gap between classical physics' singularities and the observable cosmos.

Beyond the Black Hole Universe: Exploring the Multiverse

The universe's grand narrative, as revealed by cosmology, remains largely unwritten. While the intriguing idea of our cosmos existing within a black hole challenges our current understanding, it represents just one of many vibrant avenues of inquiry at the forefront of this dynamic field. The effort to comprehend the universe's origins, its fundamental components, and its ultimate destiny is a multifaceted quest, utilizing a range of theoretical frameworks and observational strategies. Beyond the conceptual leaps required to reconcile quantum mechanics with gravity for the black hole hypothesis, cosmology is actively investigating other profound questions, each with the potential to reshape our cosmic perspective.

One of the most captivating and philosophically provocative frontiers is the concept of the multiverse. This idea, stemming from various theoretical considerations in physics, proposes that our universe may not be unique but rather one among an immeasurably vast collection of universes, each potentially possessing its own distinct physical laws, constants, and even dimensions. The notion of a multiverse arises naturally from several leading cosmological and particle physics theories, making it more than mere speculation.

Eternal Inflation and the Multiverse

Inflationary cosmology, the predominant model for the universe's earliest moments, provides fertile ground for multiverse scenarios. During the inflationary epoch—a period of rapid exponential expansion shortly after the Big Bang—quantum fluctuations are stretched to cosmic scales, seeding the large-scale structure we observe today. In many inflationary models, this process does not simply cease everywhere simultaneously. Instead, inflation might continue eternally in some regions of spacetime, while "bubbling off" new, separate universes that undergo their own Big Bangs and subsequent evolution. This scenario, known as eternal inflation, suggests a perpetually inflating spacetime that constantly spawns new pocket universes, creating an immensely diverse cosmic landscape. Our observable universe would then be just one such pocket within this unimaginably vast cosmic foam. The implications of eternal inflation are profound: if true, it would mean that virtually every possible outcome of physical processes is realized somewhere in this grander multiverse. This raises fundamental questions about the uniqueness of our physical laws and constants. Are the values we measure—the strength of gravity, the masses of fundamental particles, the cosmological constant—simply those that happen to be favorable for the existence of life, or are they governed by deeper principles that select specific configurations from an infinite array of possibilities?

Multiverse theories are supported by two primary concepts: string theory and quantum mechanics. String theory, a potential unified theory of fundamental forces, posits the existence of extra, curled-up spatial dimensions. The way these dimensions are "folded up" dictates the physical laws of our four-dimensional universe. With an estimated 10^{500} or more possible compactifications, known as the "string theory landscape," each represents a distinct vacuum state and potentially a different universe with unique properties. If these states can exist in various spacetime regions, possibly through eternal inflation, the string theory landscape provides a tangible model of a multiverse. A key challenge is understanding our universe's specific location within this landscape and identifying any observable cosmic consequences.

The Many-Worlds Interpretation (MWI) of quantum mechanics, proposed by Hugh Everett III, offers another pathway to a multiverse. MWI suggests that every quantum measurement causes the universe to split into multiple parallel branches, with each branch embodying a different possible outcome. This implies that all quantum possibilities are realized in separate, non-communicating universes. If MWI is correct, every decision and quantum event leads to a branching of reality, resulting in a continuous proliferation of universes driven by unfolding quantum probabilities.

Directly detecting a multiverse remains a significant challenge, as most multiverse scenarios involve causally disconnected universes. However, cosmologists are investigating indirect evidence. Collisions between our universe and other "bubble universes" in the past could have left detectable imprints on the Cosmic Microwave Background (CMB), appearing as specific patterns of temperature or polarization anomalies. Ongoing searches for these "cosmic bruises" in CMB data have yet to yield definitive evidence. The anthropic principle also offers a philosophical argument for a multiverse, suggesting that the fundamental constants of our universe are precisely tuned for life because we exist in a multiverse with a vast range of physical laws.

Beyond the multiverse, the enigmatic nature of dark energy presents another major frontier in cosmology. Discovered in the late 1990s through observations of distant supernovae and corroborated by studies of the CMB and large-scale structure, dark energy is the mysterious force behind the universe's accelerating expansion. Constituting roughly 70% of the universe's total energy density, its fundamental nature remains unknown.

The simplest explanation for dark energy is the cosmological constant (Λ), which represents a constant energy density inherent to spacetime. However, theoretical calculations of vacuum energy from quantum field theory are astronomically larger—by an estimated 120 orders of magnitude—than the observed cosmological constant. This "cosmological constant problem" is a significant unsolved problem in physics, indicating an incomplete understanding of vacuum energy or quantum gravity.

Alternative dark energy models are actively explored. Quintessence models propose dark energy as a dynamic, evolving scalar field that permeates spacetime. The energy density and pressure of this field would change over time, potentially leading to different cosmic expansion histories than a constant Lambda. Observational constraints on the equation of state parameter for dark energy, w, are crucial for distinguishing these models; a perfect cosmological constant has $w = -1$. Current observations are consistent with $w = -1$, but deviations could signal dynamic dark energy.

Other theories explore modifications to gravity itself to explain accelerated expansion, suggesting that general relativity may break down on cosmological scales. f(R) gravity theories, for instance, propose replacing the Einstein-Hilbert action with a more general function of the Ricci scalar, R, leading to accelerated expansion without the need for dark energy. Similarly, extra-dimensional theories, such as brane cosmology, suggest our universe is a "brane" embedded in a higher-dimensional "bulk," where gravity could leak, altering its behavior on large scales. These modified gravity theories must remain consistent with precise tests of general relativity in the solar system and binary pulsar systems while reproducing observed cosmological phenomena without introducing new problems.

The pursuit of understanding our universe is an ongoing, collaborative effort that pushes the boundaries of theoretical physics and observational astronomy. While significant progress has been made, particularly in charting celestial movements and uncovering the intricate dance of galaxies, the cosmos continues to reveal new mysteries. This enduring quest for cosmic understanding is not merely an academic pursuit; it is a fundamental aspect of the human experience.

Cosmology has been marked by continuous refinement and paradigm shifts. Discoveries like the expanding universe and the existence of dark energy and dark matter have challenged existing frameworks and spurred new theoretical and observational endeavors. The scientific process, characterized by hypothesis, experimentation, and rigorous analysis, allows us to progressively uncover the universe's complexities. Even theories that are ultimately found to be incomplete or incorrect contribute to our understanding by defining the limits of what is possible and guiding future research.

This relentless pursuit is driven by innate curiosity and a desire to decipher the fundamental rules governing the universe. It is a collaborative effort, spanning generations and continents, where theoretical physics, observational astronomy, and cutting-edge technology converge. Powerful telescopes like the Hubble and James Webb Space Telescopes, along with sophisticated ground-based observatories and particle accelerators, have provided unprecedented views of the cosmos, allowing us to probe cosmic history in remarkable detail.

Scientific discovery is inherently iterative. We build upon previous work, constantly questioning, testing, and expanding existing knowledge. Breakthroughs in one area often have ripple effects across the entire field. For instance, quantum mechanics, initially developed for subatomic particles, has profound implications for cosmology, informing theories about the early universe. Similarly, advances in gravitational wave astronomy, pioneered by the LIGO and Virgo collaborations, have opened a new window onto the universe, allowing us to detect events like the merger of black holes and neutron stars that are invisible to traditional electromagnetic telescopes. These gravitational waves offer a unique probe of gravity and spacetime.

The history of cosmology is rich with examples of audacious ideas, once considered fringe, that have become cornerstones of modern understanding. The Big Bang theory, initially met with skepticism, is now supported by overwhelming evidence, including cosmic microwave background radiation. The concept of inflation, a period of rapid expansion in the universe's earliest moments, elegantly solves several problems with the standard Big Bang model and provides a framework for understanding the origin of cosmic structure. These theories, born from bold theoretical leaps and validated by meticulous observation, underscore the power of human ingenuity in unraveling cosmic secrets.

Observational efforts to understand dark energy are intensive and diverse. Large-scale galaxy surveys such as the Dark Energy Survey (DES), the Legacy Survey of Space and Time (LSST), and the Euclid space telescope are meticulously mapping the distribution of galaxies and the growth of cosmic structure over time. By measuring galaxy correlations, weak gravitational lensing, and the expansion rate as a function of redshift, these surveys aim to precisely determine the equation of state of dark energy and its potential evolution. Baryon Acoustic Oscillations (BAO), imprinted in the distribution of matter, serve as a standard ruler, enabling cosmologists to measure distances and the expansion rate at different epochs. The increasing precision of BAO measurements provides crucial constraints on dark energy models.

Another critical area of investigation is the nature of dark matter. While not directly driving accelerated expansion, dark matter constitutes about 25% of the universe's energy density and is crucial for structure formation. Its gravitational effects are evident in galaxy rotation curves, the dynamics of galaxy clusters, and the patterns in the Cosmic Microwave Background (CMB). However, like dark energy, its fundamental composition remains elusive. Leading candidates include Weakly Interacting Massive Particles (WIMPs) and axions, but direct detection experiments have so far yielded null results, leading to increased interest in alternative candidates or entirely new theoretical frameworks. The interplay between our understanding of dark matter and dark energy is also an active area of research, as some theoretical models propose a connection or unified explanation for both phenomena.

The ongoing exploration of these cosmological frontiers—the multiverse, dark energy, dark matter, and modified gravity—is fundamentally intertwined. Progress in one area can illuminate others. For example, evidence for the multiverse might strengthen anthropic explanations of the cosmological constant. Conversely, precise measurements of dark energy's properties could rule out many modified gravity theories, indirectly supporting a cosmological constant or hinting at new physics. Each new dataset and refined theoretical model brings us closer to a more complete and coherent picture of the cosmos and our place within it. The challenges are immense, but the potential rewards—a deeper understanding of reality itself—are immeasurable. The universe, it seems, still holds many secrets waiting to be uncovered.

The field of cosmology is on the verge of groundbreaking discoveries. The mysteries of dark energy and dark matter, which account for roughly 95% of the universe's energy, are paramount challenges in physics, suggesting the existence of new particles, forces, or modifications to gravity.

The perplexing nature of dark energy, the force behind the universe's accelerating expansion, remains a central question. Is it a cosmological constant, a remnant of the early universe's vacuum energy, or a dynamic field representing a new fundamental force? Future surveys like the Legacy Survey of Space and Time (LSST) and the Euclid space telescope will be vital in discerning these possibilities by meticulously mapping the universe's expansion history.

Equally pressing is the quest to understand dark matter. Despite observing its gravitational effects, its constituent particles have eluded direct detection. Underground experiments are actively searching for Weakly Interacting Massive Particles (WIMPs) or axions. The absence of these detections is prompting cosmologists to explore broader possibilities, including alternative interaction mechanisms or entirely new theoretical frameworks. The eventual discovery of the dark matter particle would revolutionize our understanding of fundamental physics and cosmic composition.

Furthermore, unifying gravity with quantum mechanics, the holy grail of theoretical physics, is intrinsically linked to cosmology. Theories like string theory or loop quantum gravity are expected to shed light on the universe's initial moments, the nature of black hole singularities, and the very origins of spacetime. These theories offer glimpses into realms where current physics breaks down, hinting at entirely new principles. The intriguing concept of a multiverse, suggested by some inflation and string theory models, adds another layer of complexity, with indirect observational signatures actively being sought in the cosmic microwave background.

The spirit of cosmological exploration transcends theory and observation, encompassing our approach to scientific inquiry. It demands embracing uncertainty, formulating bold hypotheses, and rigorously testing them against evidence. This relentless pursuit, driven by human resilience and intellectual curiosity, pushes the boundaries of our understanding despite immense challenges and profound unknowns. The universe, an endless source of inspiration, fuels a continuous journey of discovery, where each revelation unveils new vistas and prompts further questions.

As we stand on the cusp of potential breakthroughs in understanding dark energy, dark matter, and fundamental physics, it's crucial to acknowledge that the universe will likely continue to surprise us. The history of science teaches us that our current models are merely approximations, glimpses of a deeper reality. The act of seeking knowledge, engaging with the cosmos through observation, theory, and experimentation, is a profoundly enriching experience, connecting us to something far grander than ourselves, fostering both humility and empowerment.

The enduring quest for cosmic understanding reflects our innate desire to comprehend our existence and place within the universe's grand narrative. This journey has no foreseeable end; unraveling one mystery often reveals many more. The beauty of cosmology lies not only in its answers but in the questions it inspires and the intellectual adventure it offers. It serves as a constant reminder of the universe's infinite possibilities and that the pursuit of knowledge is among humanity's most noble and rewarding endeavors. The universe holds countless wonders awaiting discovery, and the journey of exploration, with its challenges and triumphs, is its own profound reward, continually reshaping our perception of reality and our understanding of what it means to be part of this magnificent cosmos.